Snowstruck

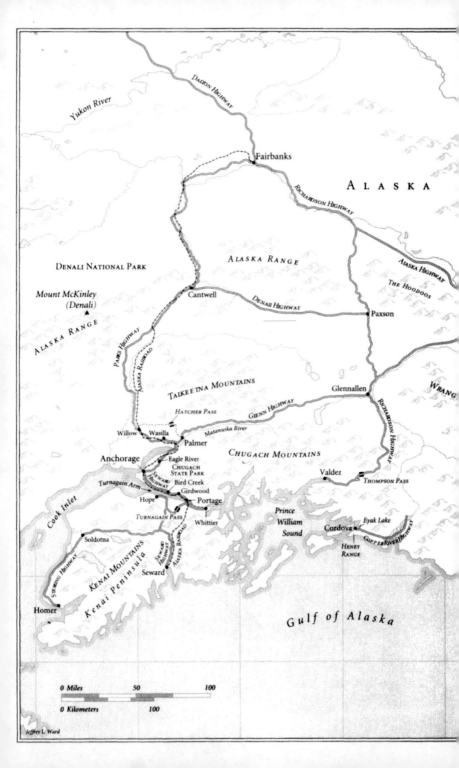

"*Snowstruck* should be required reading for any snow enthusiast, not only for its eloquent passages on powder blasts, slab avalanches, and shifting water-vapor molecules, but because it takes avalanches out of the statistical realm and into the human one."
—*Skiing Magazine*

"[A] chilling cocktail of adventure, science, and memoir."
—*Best Life Magazine*

"Captivating . . . Readers will be enthralled by the author's revelations. This page-turner should be required reading in germane geographical areas and is highly recommended."
—*Library Journal* (starred review)

"This book is an electrifying account of the dangers of avalanches, their causes, their victims, and—thanks to Fredston—sometimes their victims' rescue."
—*Booklist*

"Fredston brings passion and a wealth of experience to her story. . . . An informative and powerful cautionary lesson."
—*Kirkus Reviews*

"Jill Fredston is on frighteningly intimate terms with the grandest and most violent tantrums of snow thrown by our planet. *Snowstruck* abounds with wild, heartbreaking, unforgettable stories of human striving and fate. We are front and center at Fredston's never-ending duel with avalanches and fortunate to have such a dazzling writer placing us there."
—Howard Norman, author of *The Bird Artist* and *In Fond Remembrance of Me*

"*Snowstruck* is a thrilling, fascinating, and inspiring narrative about avalanches: their causes, their victims, and their predictability. Fredston draws the reader into the story with her amazing real-life experience. It's a book you can't put down. One of the best I've read in a long time." —Lynne Cox, author of *Swimming to Antarctica*

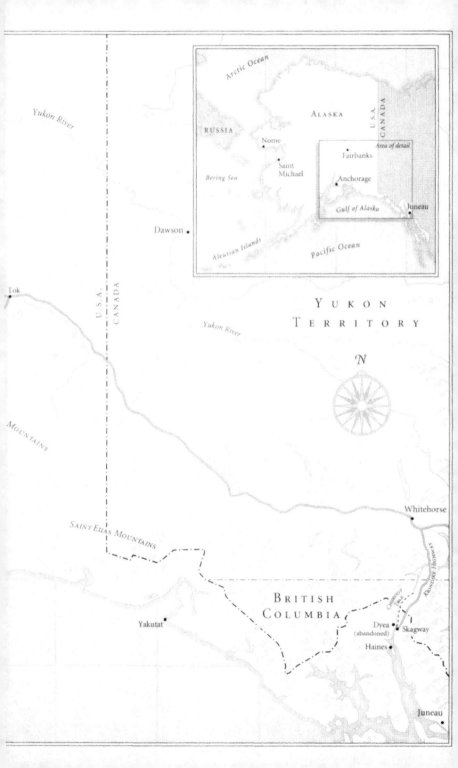

Jill Fredston

Snowstruck

In the Grip of Avalanches

A HARVEST BOOK
HARCOURT, INC.

Orlando   Austin   New York   San Diego   Toronto   London

For Doug Fesler, of course

———————————

www.HarcourtBooks.com

The Library of Congress has cataloged the hardcover edition as follows:
Fredston, Jill A.
Snowstruck: in the grip of avalanches/Jill Fredston.—1st ed.
p.   cm.
1. Avalanches.   2. Natural disasters.   3. Survival skills.   I. Title.
QC929.A8.F74   2005
551.3'07—dc22   2005020454
ISBN-13: 978-0-15-101249-7   ISBN-10: 0-15-101249-0
ISBN-13: 978-0-15-603254-4 (pbk.)   ISBN-10: 0-15-603254-6 (pbk.)

Text set in Minion
Map on pages iv and v by Jeffrey L. Ward
Designed by Linda Lockowitz

Printed in the United States of America

First Harvest edition 2007
K J I H G F E D C B A

Title page photograph © Chuck O'Leary Collection

Contents

Snowstruck

CHAPTER 1
Moments of Truth

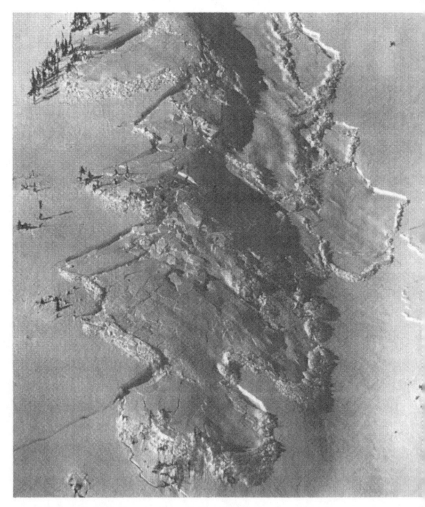

A slab avalanche about two seconds into its release © JIM BAY

JERRY HAD BEEN SHOVELING FOR DAYS. IT FELT LIKE MUD, but it was only snow, snow, and more snow. His town was more a battleground than the dreamy winter scene of paperweight globes. Trees were toppling, roofs were in danger, and telephone poles plastered with wind-driven snow appeared double their normal diameter. Most mornings the lean fifty-year-old flooring contractor and occasional longshoreman didn't bother to throw on more than flannel slippers and bathrobe to shovel a soggy furrow from house to car. But this January 2000 morning in the isolated fishing port of Cordova, Alaska, the monitor heater in his house was down and Jerry had burned the last of the firewood stored close at hand, so he told Martha, his partner of more than twenty years, that he would dig his way out to the detached garage for another load. Knowing this wouldn't be a trivial project, he pulled on gray sweatpants and a hooded shirt, ribbed cotton socks and shoes.

Sequestered on the rocky eastern edge of Prince William Sound in southcentral Alaska, Cordova is not connected to the rest of the state by road. Instead, the twenty-five hundred year-round residents of Cordova depend on the Copper River Highway—which might otherwise seem a misnomer for a two-lane road with intermittent traffic—as a vital corridor linking the town at waterside with the airport twelve miles east. If you want to leave Cordova, you must do so by air or water. If you want to stay, you can forget about traffic lights or parking meters and take your pick of seven bars. Jerry and Martha, who moved north from Virginia in 1980, were among those who stayed. Since 1991 they had lived in what seemed an idyllic spot, sandwiched between the crystalline emerald water of Eyak Lake and the steep slant of the Heney Range. Known as 5.5 Mile for its distance from town, the neighborhood consisted of just over a dozen houses and two warehouses strung along a short dirt road that looped at both ends to join the highway. Some of the dwellings, particularly those on the lakeside, were nestled in clusters of tall thick-boughed spruce, and many, in familiar Alaska fashion, were surrounded by an array of boats on trailers, storage sheds, motor homes, and long-dead vehicles scattered about like lawn ornaments. Just above the houses, the mountain's slope was a clean white scar in winter and a jungle of prickly bushes in summer.

Jerry had been making tedious progress when he heard a rumbling like distant thunder. Looking up the broad fan-shaped lower slope on the far side of the Copper River Highway, he saw a roiling, fast-falling curtain of snow. Jerry had lived at the foot of the mountain long enough to be acquainted with avalanches. The one that had grazed the neighborhood a few years earlier had been nothing more than a wraith of powdery snow and wind by the time it reached Jerry's doorstep, and Jerry figured this one would be the same. Glum at the prospect of even more shoveling, he hurried inside his house. He had one hand on the royal-blue boiler-room door that would lead him upstairs to Martha, who would be lingering over coffee and the chat of morning television, when he heard the roar that would change his life.

Almost before he could register alarm, Jerry pivoted and saw the back wall of his house rushing toward him as if it had been shot from a cannon. In a split second, he guessed, he'd be pinned against the next wall and that would be the end of him. Then that wall exploded in a starburst as well, and Jerry became just another piece of flying debris in a mad disintegrating whirl of blue door, furnace, hot water heater, wooden beams, insulation, fractured pipes, Sheetrock, and arrows of shattered glass. Somewhere in the same maelstrom was Martha.

❋ A FEW HOUSES AWAY, twenty-four-year-old Christal Czarnecki had, as always, risen early enough to walk the dogs before work, leaving her boyfriend, Wes Burton, in bed nursing an injured knee. She was accustomed to shuffling along the loop road through heavy snow, her head bowed in deference to gusty winds—Cordova could credibly claim patent to virulent storms. Christal had, however, no presentiment that she was following the footsteps of another woman who had been out for a similar walk thirty-six years earlier when an avalanche billowed across the road, burying her to the waist and killing her dog. Still, Christal had forged only a hundred yards, barely able to see her feet in a whiteout thickened by fifty-mile-per-hour winds, before she decided to cut her walk short and turned for home.

From bed, the place beside him crinkled and cool, Wes heard Christal start the shower in the adjacent bathroom. He also heard a distant hum he assumed was a road grader rattling along the highway. Within seconds, though, the noise swelled to a rumble, as if a tractor was idling impatiently outside the house. Wes's last thought before events became visceral was that a jet was flying directly overhead and at unusually low altitude.

Without a warning flicker, the power cut off. So did the water. Christal, naked in the windowless bathroom, felt the house tremble like the flame of a candle, and then shudder. As she scrambled for balance, the shower door derailed and clattered against the far wall while shelves and picture hooks rained objects. The crashes confirmed what Christal knew but could not comprehend: the house was moving.

The ruckus was abruptly replaced by an even eerier stillness. Wet and shivering, Christal crept barefoot from the dark bathroom. Wes met her at the door, his muscled arms outstretched. Together, as though they were venturing onto a new planet, they explored their house. In the kitchen the refrigerator lay tipped, its door ajar and contents spilled in a slurry. Next to it the microwave sat on the floor, jauntily topped by a lamp shade. Drawers that had rolled open brimmed with escaping utensils. It looked as though everything in the room had been briefly granted life and then caught in the middle of a mad orgy.

Back in the bedroom, Wes and Christal stood in the jagged opening where moments before they could have stepped through a sliding glass door onto a balcony affording a grand view of Eyak Lake. But the balcony was gone; later they would find it cast adrift in the snow. Only when Wes and Christal saw the roof of their Toyota RAV4 inches below their toes did they realize that the carport and workshop that constituted the first floor of the house had disintegrated. Untethered, the second floor had sailed twenty feet through the air like a raft, with Wes and Christal aboard. Still more or less intact, the second floor had parked itself on the hood of the Toyota, its leading edge rammed against the steering wheel.

❄ WHILE JERRY AND his neighbors had spent many of the days leading up to January 26, 2000, shoveling nearly six feet of snow at sea level, the mountain that shadowed them had spent most of the month in the clouds, unobtrusively loading itself with twice as much snow and weathering hurricane-force winds. The huge treeless bowl just below the summit ridge curved like an amphitheater equal in surface area to a large sports stadium; it could not have been engineered to collect new or wind-whipped snow more efficiently. Even the tallest professional basketball players would have disappeared altogether in the deeper drifts.

January 26 had dawned another white day. If the top of the mountain hadn't still been shrouded, witnesses might have seen a thin crack arc across the slope, announcing itself with a boom that was likely lost in the ratcheting wind. At first the crack would have looked like a tear in the clouds, a rent easily stitched up with white thread and forgotten. But, in a blink, it would have wrapped more than a half mile around the bowl and, as if yawning, widened as a plate of snow more than six feet thick lost its grip on the mountainside. More ruptures would have shot along the sides, through the middle, and along the bottom edge of the plate until, for one suspended moment, it resembled a child's jigsaw puzzle. And then there would have been too much motion for the eye to track, the pieces buckling and breaking into smaller and smaller shards, the whole slope a shattering windowpane. Falling and tumbling, shoving forward a tongue of displaced air, gathering speed, the avalanche would have vacuumed up the massive quantities of loose snow that lay in its path, building a thundering cloud of its own that ballooned higher and higher into the sky, becoming the only view. This boiling cloud would have looked soft and cottony, its advancing edge dancing and surging like an ocean wave, but it was a barrage of stinging icy particles, a vortex of loud punishing wind. Behind the cloud, bumping along the craggy ground, was an unstoppable torrent of millions of pounds of snow.

The term *avalanche* stems from the French verb *avaler,* which means "to swallow" but originally meant "to descend." Within four seconds the 5.5 Mile avalanche was dropping at freeway speeds, still accelerating as it pounded down even steeper slopes and flew over the edges of cliffs. The avalanche had already plunged almost the height of the Empire State Building when it reached the narrow squeezing gut in its hourglass-shaped run. There it erupted, the cloud shinnying up the gully walls and shooting more than four hundred feet into the air. Acres of trees two or three feet in diameter that had stood for nearly a century snapped like raw spaghetti and shredded into strands almost as thin. A behemoth spruce, six feet thick and four hundred years old, sacrificed limbs. Inside the turbulent, deafening snow cloud now were hurling branches and sheared tree trunks, errant missiles seeking targets.

Jerry didn't see the avalanche until it was about twenty seconds and almost two thousand vertical feet into its life, on the last steep drop above the Copper River Highway. It didn't fall so much as pounce upon the road, obliterating nearly a thousand linear feet of pavement under piles of snow as deep as thirty feet. Its forward edge now smoking 90 to 120 miles per hour, the avalanche tore through the subdivision as though the houses were made of paper, dissipating the last of its energy only when it was more than a quarter mile out onto the ice of Eyak Lake. From start to finish, the avalanche had taken less than thirty seconds.

✳ EMERGENCY DISPATCH in Cordova is not a long bank of telephones manned by an anonymous army in an isolated soundproof room. On the morning of January 26, it was two women named Dixie and Gerry sitting in a nondescript office inside the joint police and fire headquarters, not far from the docked fishing fleet and the town's only bakery. Their early morning entries are neatly printed; clearly the available minutes exceeded the action. Most of the calls reported fallen trees, power outages, or poor visibility on the roads. But after the first 911 call was logged at 9:56 A.M., the letters begin to slant, with greater space between them, as though each is vainly trying to keep up with the next.

The first 911 call was from a 5.5 Mile resident. The second was from a plow driver who had been heading into town on the Copper River Highway. The air was still gauzy with suspended snow particles when he noticed a large spruce tree across the road. It was not until he slowed that he was able to distinguish the mound of ruffled snow blocking his path. Unseen, death had darted in front of his plow.

Ducking into the hallway, Gerry alerted Bob Plumb, the deputy fire chief and the sole paid fire department employee, that "something big" had happened at 5.5 Mile. Plumb, already dressed to shovel the headquarters' roof, instructed, "Call Dewey!" and ran for his truck. Three minutes later dispatch logged the first damage report from the plow driver: "Gunnerson, Jerry LeMaster, Clayton, Logan houses hit hard—looking for survivors."

Plumb turned off the highway onto the loop road leading to the 5.5 Mile houses, but he drove only scant yards before the road disappeared under a pile of snow and a tangle of fractured trees. Ahead of him, little appeared to be where it should. Several houses, including the one Wes and Christal rented, had been blasted off their foundations. They lay at odd angles, mortally wounded, with windows blown out, trees piercing their walls like javelins, and roofs torn off. One warehouse had disappeared altogether, its metal roof crumpled like toilet paper around the remains of some nearby trees. "Hello!" Plumb yelled. "Is anyone here?" People began to emerge through the haze. Plumb, numbed by the immensity of the avalanche, didn't yet realize that he was seeing only one edge of the destruction.

Before his pager shrilled, Dewey Whetsell had been preoccupied, like almost everyone else in town, with shoveling his roof. Retired after twenty-eight years of active firefighting duty and still serving as Cordova's elected volunteer fire chief, Whetsell was the heart of the fire department. He'd spent a lifetime responding to shipboard fires, blazing houses, car wrecks, medical emergencies, underwater accidents, and other disasters. A man of average height, he has an expansive forehead furrowed from decades of making decisions with consequences, a lookout tower of a nose that supports large glasses, and thin hair that clings to the outlying reaches of his head and jaw.

Five minutes after the page, Whetsell pulled onto the blocked loop road wearing a fire jacket marked CHIEF and a white hardhat. Assuming his customary role as incident commander, he asked Plumb to set up a staging area for the rescuers who would soon flood in from town while he sized up the damage and took stock of who was missing. Whetsell then began scrambling over snow that felt more like concrete than the mush he had been about to shovel at home. As he picked his way across the tops of buried cars, broken trees, and the sharp protruding bits of buildings, the scene took on a new level of detail. A fishing boat had surfed into a concrete foundation; a washing machine perched on the snow's surface, pounded down to the size of a toilet; a sink hung twenty feet above his head in the branches of a spruce tree. Even three hundred feet from shore, the lake ice was littered with fiberglass insulation and furniture. Through an odd mix of pungent smells came the unmistakable odor of leaking fuel oil.

As Whetsell penetrated beyond Plumb's farthest point of exploration, he came upon a knot of people in what appeared to be the hardest-hit section of the subdivision—the semicircle of land between the highway and the loop road. One in the group was the plow driver, who had approached on foot from the airport side of the avalanche. Another was 5.5 Mile resident Kevin Clayton, a retired police chief who had jumped back to work trying to make sense of the chaos. Whetsell, disoriented, asked where Clayton's house was. Clayton pointed to a green building that Whetsell hadn't recognized; he remembered Clayton's home as having two stories, not one. Only then did he register that he was looking at the upper floor of the house, which had been shoved forward and now drooped, as though exhausted, onto the ground. Whetsell knew that his friends Jerry LeMaster and Martha Quales lived next door to the Claytons—he'd been to their place often—but he didn't see the damage there that he saw among the other houses. LeMaster's house wasn't off its foundation with a missing first floor, punctured walls, and trees resting on the roof. It was gone.

Kirk Gunnerson lived on the other side of Jerry and Martha. He had been upstairs chatting on a cordless phone when a blast of wind like a freight train preceded what felt like a bomb blast, interrupting Gunnerson midsentence and sending him swimming for his life through a turbulent stream of snow and disintegrating house. Seconds later he was spewed aside and left standing in his underwear, phone still in hand. Anesthetized by shock, Gunnerson stepped from the wreckage through a hole in the roof, dialed his employer, and reported that he wouldn't be coming to work that day.

In the opaque predawn hours, Cordova's superintendent of schools, who lived at 5.5 Mile, had had to decide whether to cancel classes. The sheer volume of new snow would have paralyzed most communities, but Cordovans cannot afford to set aside daily life for as common an inconvenience as weather. The schools had stayed open, a decision that inadvertently saved lives among the students, parents, and staff who lived at 5.5 Mile. Running late, the school bus had rolled through the neighborhood barely an hour before the avalanche. Of the residents whose homes had been hit, Kirk Gunnerson, Wes Burton, Christal Czarnecki, and Kevin Clayton were accounted for. But how many others were missing?

Whetsell surveyed the scene. If anyone was trapped in the wreckage under his feet or buried in the heap of snow on the highway, it would be a race to find them before they ran out of air. Despite his proficiency in handling crises, he felt as though he were grinding gears. One minute his mind spun with the need to formulate a plan; the next he felt stupefied, unable to get his bearings, dulled into inaction by the magnitude of the damage. The equivalent of a city block had been obliterated.

Cordovans rose to the task of figuring out who was missing. Every few minutes the local radio station, KLAM, announced that 5.5 Mile residents should check in with police and urged others to verify the whereabouts of anyone who might have been driving the Copper River Highway at the time of the avalanche. People called schools and workplaces; friends combed favorite haunts; family members tracked one another down. The dispatch log is a litany of anxiety. At 10:10 Kathy Sjostedt reported that her husband's twin brother, Don, had been at their warehouse, which the dispatcher knew had been reduced to rubble. At 10:12 dispatch radioed Bob Plumb telling him to watch for Sjostedt and his truck. "Okay," Plumb responded, thinking it a hopeless task since he could see the warehouse's metal roofing panels out on the lake ice. But by 10:23 Sjostedt had been accounted for—minutes before the avalanche, his wife had summoned him home when a tree landed on their house. There was concern as well for a man who had been driving the Copper River Highway in his green Dodge pickup. He was found twenty minutes later among the volunteer searchers. In this fashion, names flitted on and off the log sheets until only two names lingered: Jerry LeMaster and Martha Quales.

Pagers beeped all around town, summoning volunteer firefighters, medics, and members of the local ski patrol. Cordovans are a self-reliant breed. In 1989, when the Exxon *Valdez* impaled itself on rocks not far from Cordova, dumping more than 11 million gallons of crude oil into Prince William Sound, Cordovans didn't wait for cleanup crews but jumped into their own boats to try to contain the spill. Now, as word of the avalanche spread, dozens of volunteers, most wearing brightly colored foul-weather fishing gear and scuffed rubber boots, streamed to the fire station and were dispatched in small teams to 5.5 Mile.

Although Jerry owned a Harley-Davidson motorcycle and thought of himself as a biker, he and Martha were most often seen hauling grandkids around in their brown 1992 Chevrolet Astro van. Martha's daughter informed dispatch that if either Jerry or Martha had ventured out that morning, they had almost certainly driven the Astro. While police scouted the streets and marina for the van, Whetsell asked the first rescuers at 5.5 Mile to push makeshift probes of aluminum electrical conduit into the snow in the area that Kevin Clayton guessed was Jerry and Martha's driveway. By 11:54 they struck metal and confirmed that the van had been parked. Probing and digging proved the van empty.

Whetsell, still grasping for a plan, moved everyone about thirty feet from the house's original location, where neighbors were beginning to identify pieces of wreckage as Jerry and Martha's belongings. In the hope that Jerry, a fifteen-year veteran of the volunteer fire department, had been wearing his pager when the avalanche hit, Whetsell ordered that all pagers, radios, and cell phones be switched off. A distressing silence ensued, given dimension only by the wind and the rustle of agitated rescuers attempting to be still. When Whetsell gave dispatch the go-ahead to issue the fire department page, called a "tone-out" in the trade, the assembly of firefighters, police, residents, mountaineers, and equipment operators knelt, eyes squinted in concentration, listening hard. But the more intensely the group listened, the more amplified the silence. Beseeching faces swiveled toward Whetsell as though he should know exactly what to do next.

Probing was futile—the conduit kept dead-ending into chunks of wall. All Whetsell could do was direct searchers to dig.

Whetsell and Plumb were both nagged by the same cutting fear. Less than a year earlier, the fire department had led rescuers into a dark canyon on the opposite side of Eyak Lake, where a hydroelectric project was under construction and an avalanche had buried an equipment operator. In winter this Power Creek site is a horribly dangerous place, equivalent to working at the bottom of a funnel. The canyon's 55-degree slopes circumscribe the view such that less than a third of the 3,000-foot mountain can be seen from the ravine floor. Avalanches arrive as sneak attacks, routinely vaulting off the mountain and plunging into Power Creek. Occasionally they have enough momentum to charge two hundred feet up the timbered slope on the far side of the valley. Sometimes, as during that accident, avalanches dam the river and cause it to flood.

At Power Creek only luck had saved Plumb from breaking the cardinal rule of rescue, which is not to make a bad situation worse by killing more people. When a second and then a third avalanche cascaded into the river, depositing more than fifteen feet of snow on top of the first slide, several firefighters found their boots mired in the wet debris as if in quicksand. They escaped only by wrenching their feet free and running in socks across the moving snow. One woman slipped through a chasm and into the clutch of the rising river but was tugged to safety.

At 5.5 Mile the potential consequences of another avalanche were worse than at Power Creek because more rescuers were on site and they were dispersed over a wider area. Without enough gear to go around, most searchers weren't wearing avalanche rescue beacons, small electronic devices emitting signals that would afford them the best chance of being found if trapped by a second avalanche. The continuing storm prevented Plumb and Whetsell from seeing how much snow still threatened from above. Once they made the decision to search, they could do little other than keep strict account of who was in the area and post a guard armed with a compressed air horn to warn rescuers if he saw another slide charging out of the clouds—a gesture that could be compared to waving a flag from beneath the hooves of a stampeding bull.

In backcountry avalanche accidents, the usual tools for finding buried victims are shovels, avalanche beacons, and eight-foot-long probes not much thicker than a pencil. But avalanche rescue in settled areas is an entirely different animal. People don't wear transmitting beacons in their living rooms, and probes are ineffectual because so much solid material is mixed into the snow debris. That leaves heavy equipment, shovels, and a variety of hardware such as chain saws, sawsalls, pry bars, and tow chains.

Trying to create a more orderly process, Whetsell established work zones. He commandeered loaders and backhoes to dig from either end of the loop road toward Jerry and Martha's house. When they had muscled to within fifty feet of each other in the area Whetsell dubbed Zone 2, a safety officer choreographed their movements and spotters scrutinized the loader buckets and cut banks as the operators made careful scoops. Zone 1, oxymoronically nicknamed the "hot zone," was the area of the house itself, where searchers wielding chain saws and shovels pried the snow-encased rubble apart piece by piece. Zone 3, where probers worked from the lakeshore up to the loop road, seemed impossibly distant, and yet it was conceivable that Jerry or Martha had been blown as far as some of their possessions, which lay as much as half a mile from the original house site. The "rehab sector," set up near the fringe of the avalanche, became the receiving area for water, hot drinks, doughnuts, and sandwiches, a place of relative quiet where rescuers could refuel.

Around 12:15, almost three hours after the avalanche, a shout of "Over here!" gathered a cluster of searchers in the hot zone. A digger had uncovered Martha's hand and arm. Both were the dark blue of only slightly faded denim. When a medic knelt to feel a pulse, allowing extra time for the slowed heart rate of a very cold person, he could detect no sign of life. It took twenty minutes to chip and pry Martha's body free. She was sitting in her living-room recliner, the television remote control still clutched in her fingers.

At 12:32 the dispatch log records Martha as 10-79, emergency operations code for dead. Saying little, the rescuers picked up their shovels and drifted back to work.

❄ THE AVALANCHE became personal for me around eleven in the morning. The call came while I was on the other phone line with Aedene Arthur, the mother of a snowmobiler who had been killed, with five other men, in a large avalanche the previous spring. She and I weren't breaking new ground. Though she often seemed to hate me for my responses, she had asked me the same questions nearly every week in the ensuing 311 days. Did I think her son had triggered the avalanche? Why had it taken two months to find him? Had he died instantly? When I told her that I had to run, that there had been an avalanche in Cordova, she gasped, the new crisis a fresh stab to a wound that refused to heal.

The call was from Sergeant Paul Burke, the Alaska State Troopers coordinator for search and rescue. All Paul asked was whether I could go to Cordova. He didn't have to explain that he needed me to assess whether the site was safe for rescuers. I work from home, so as we spoke, I had one hand inside the bedroom closet, grabbing for clothes that would keep me dry. In winter my rescue pack, with a dented aluminum shovel buckled on the outside, stands ready against the garage wall. Throwing it over one shoulder and hoisting a hockey bag filled with avalanche probes and rescue beacons, I was out the door within ten minutes of the call.

I argued with myself all the way to the airport. Actually, it was a full-fledged fight. Winds gusting over a hundred miles per hour and poor visibility had prevented small planes from reaching Cordova for three days. If the weather had improved at all, it was by the most negligible of margins. Yet I was about to climb into a plane slightly bigger than a tin can, with a pilot I didn't know. Why take the risk for someone who was almost certainly dead? After thirty minutes of burial, an avalanche victim has less than a 50 percent chance of survival. After an hour, the odds are down to a dismal 23 percent. By the time I reached Cordova—if I reached Cordova—the avalanche would be nearly four hours old.

I knew about the statistical anomalies, like the Colorado miner who spent twenty-two hours in 1986 clawing through thirty feet of snow with his hands, only to be hit by two more slides before he ultimately found safety. Another Colorado man, a highway worker, had tunneled for eighteen hours to free himself in March 1992, and his declaration that he wouldn't be alive if he had quit after seventeen and a half hours still reverberated within the walls of my mind. Though she lost a leg to frostbite, twenty-two-year-old Anna

Conrad survived five days (115 hours) of burial in 1982 in the wreckage of a large steel-girdered A-frame building demolished at the base of Alpine Meadows Ski Area in northern California. Among the most extraordinary survival tales was that of two adult sisters and an eleven-year-old girl in northern Italy in 1755. Their house, which was one of thirty destroyed in the tiny mountain hamlet of Bergemoletto by an avalanche that took the lives of twenty-two people, was crushed beneath forty-five feet of snow. They were trapped for thirty-seven days in a fetid tar-black airspace roughly twelve feet long, eight feet wide, and five feet high. A five-year-old boy with the trio died, as did the cattle, donkey, and hens that were sharing their airspace, but the milk of two hardy goats sustained them. One of the goats even gave birth a few weeks into the ordeal. The women had long been given up for dead when their brother dreamed that they were pleading for help and mustered another rescue effort.

For me, though, these were only stories. In my eighteen-year avalanche career, which began the same year Anna Conrad was buried at Alpine Meadows, I had chiseled dozens of bodies from avalanche debris and had never dug a single person out alive.

❄ ON MY WAY OUT of the house, I left a scrawled one-line message on the kitchen counter for my husband, Doug Fesler. This had long been our means of giving each other an idea of where we had run off to on scant notice, leaving the dog and a ransacked house behind. The spring before, we'd created a slapdash pile of destinations—Eureka, Cordova, Cantwell—during a bleary six-week siege that pulled us in opposite directions as we scrambled to seven separate avalanche accidents in which thirteen people died.

Doug had been gone since five that morning. On top of thick polypropylene long johns, he'd pulled on patched Gore-Tex coveralls, a red parka, and insulated rubber boots. He left the house carrying a shovel, binoculars, a notebook, a set of chains for our rusty diesel-powered four-wheel-drive Chevrolet Suburban, and thirty years of experience evaluating potential avalanche danger. It was still dark as he rattled the forty-five miles down the Seward Highway from the mountains above Anchorage where we live to Portage, so-named because the snowy U-shaped pass at the end of the steep-walled valley was a traditional portage to Prince William Sound.

When the wind isn't blowing strong enough to rip car doors off their hinges, Portage is a Shangri-la of sculpted peaks, fissured glaciers, waterfalls, and a lake punctuated with icebergs. As an exigency of World War II, two and a half miles of railroad tunnel were blasted through two mountains, connecting Portage to an ice-free port in western Prince William Sound so that supplies brought in by ship could be shunted inland. At the port end of the tunnel is Whittier, where most of the two hundred residents live in a stunningly ugly fourteen-story concrete bunkerlike tower constructed by the U.S. Army. Few know it, but during World War II, Whittier was designated an alternative White House site, a refuge to which the president could be whisked if Washington were attacked. Rimmed by mountains with hanging blue walls of ice, and more often than not swaddled in clouds that drop roughly fifteen feet of rain and twenty feet of snow annually, Whittier has formidable natural defenses. Like Cordova, it is prone to receiving ungodly amounts of precipitation in short periods. One storm dumped seven inches of snow every hour for sixteen straight hours, smothering the town in more than nine feet of new snow. Nor is it unusual to have twelve feet of dense accumulation on the ground at sea level in May, burying some buildings to their second-floor windows. In winter, which encroaches upon most of the school year, many of the thirty or so Whittier children reach school via an underground passage.

At Portage Doug pulled up behind half a dozen rigs idling at the entrance to the railroad tunnel, most with construction workers wedged behind their steering wheels. Crews were working to convert the tunnel into a corridor that could also be used by passenger cars, tour buses, and trucks, a controversial project because of the easy access it would afford to relatively remote and wildlife-rich Prince William Sound. Doug had been hired as a consultant a week earlier at the insistence of the electrical linemen who had been installing a transformer outside the Whittier tunnel entrance when an avalanche ambushed them. The slide blew three 200-pound men a hundred feet through the air and turned a 6,000-pound equipment trailer on its side. The linemen had refused to go back to work without an avalanche forecaster present to assess the hazard.

When you come out of the tunnel on the Whittier side, you are immediately in the line of fire of big avalanches that can drop from as much as three thousand feet above you. It's as though you poke your head out of a manhole and find yourself in the middle of a busy street, with traffic whizzing all around. On Doug's first day, he accompanied the crew through the tunnel to the Whittier side and emerged into a raging blizzard. "The snow was falling like baseballs," he recalls. The wind was blowing sixty to seventy miles per hour, with gusts to eighty. It was impossible to see more than twenty feet beyond the tunnel's portal. "These big tough guys who I'd never met before and who didn't know me from Adam were looking at me with distrust written all over their faces," Doug says. "I knew they were thinking, *This idiot thinks it's safe to work in this?*"

What the workers didn't yet know was that just before the storm, Doug had gone up in a helicopter to land on the ridges above the job site and lowered himself by rope into the starting zones of the avalanche paths. When he'd dug test pits, he found the layered snowpack to be stable, with the exception of a skim of powder on the surface. The winds were stripping that surface snow off the slopes, so the crew was in no immediate danger.

Now, a few days after beginning work, Doug again waited with the construction crews. Only diesel vehicles were allowed through the single-lane tunnel, which was opened at 6:00 A.M. for the workers traveling toward Whittier, and then again in late afternoon for their return to Portage. Anyone who missed these windows was free to don a hardhat and ride a bicycle through the poorly lit, dripping passage.

When Doug emerged from the tunnel, he saw stars; Whittier was enjoying a rare clear sky. But as the day lightened, he could see a band of black nasty clouds to the east, in the direction of Cordova. At 8:30 A.M. he told the linemen that a storm was advancing, but that with luck they'd squeak out a day's work. By 9:00 A.M. it was snowing. By 10:00 it was snowing in earnest. By 11:00 six inches of new snow was on the ground and the winds were starting to blow. By noon Doug warned the electrical crew's foreman that they would likely have to quit early.

———

✳ I COULDN'T REACH Cordova without flying virtually over Doug's head, through the deteriorating weather that was beginning to make him feel more and more like a nervous caged lion. At the airport I asked a few perfunctory questions about the instrumentation and whether we had enough fuel to return to Anchorage if landing in Cordova proved impossible. But I barely registered the answers as I strode toward the tarmac, exchanging sketchy details about the accident with the four other passengers. I was filling the extra seat on a plane chartered by the *Anchorage Daily News* and a local television station. Introductions weren't necessary; I already knew the reporters, photographer, and cameraman shoving their equipment aboard. I thought of them as disaster dates, since we tended to see each other during the media blitzes sparked by emergencies.

We took off around noon and had been airborne little more than a minute when sixty-mile-per-hour winds began to kick us around like a soccer ball. Leaden clouds folded in around the wings, pasting the windows with snowflakes. When particularly violent gusts hit, my stomach lurched into my throat while my skull rose to punch the ceiling. I considered pulling the life vest from under the seat to protect my head and made fervent promises never to be so stupid again. Though I was sitting knee to knee with two of the journalists, I tried not to look them in the eyes. Their fear was so obvious that it made mine feel more real.

As I stared out the window, seeing nothing, I couldn't restrain my mind from remembering the last time I'd made promises not to fly in such weather. Fifteen years earlier I'd been in an HH3, a cavernous camouflage-green helicopter flown by the air force. We left Anchorage in the sunshine but blundered into the same blizzard that had caused the

C-135 military plane we were looking for to crash in the glaciated take-no-prisoners mountains rimming Prince William Sound. Following the coastline, bellying close to the water, we had little margin for error when, abruptly, the gray ocean became indistinguishable from the sky. I was not wearing a headset, so I couldn't hear the pilot yell over the intercom to his navigator, "Tell me which way the shore is going to turn! Tell me what's coming up!" The rescuer on the mesh bench across from me heard every word though, and began motioning to me and the rest of the mountain rescue team to cinch our shoulder harnesses and tighten our helmet straps. Directly ahead a wall of trees emerged from the murk; the coast was curving opposite the way the navigator had said it would. With undisguised panic, the navigator shouted, "HARD RIGHT! HARD RIGHT NOW!" and the pilot banked the helicopter so steeply that it keeled over like a capsizing sailboat. I didn't need to hear the shrill stall alarm over the intercom to know that we were about to crash; I could see tree branches whipping by the windows, little more than arm's length away. Then the clouds cleaved open fortuitously, allowing us to escape into blue sky. We spent the next four days searching, even refueling from a tanker in the air so that we could hunt for thirteen nauseating hours at a stretch, but we did not find the plane.

As spring burgeoned months later, the plane's thirty-foot tail section melted into view on the same mountainside we'd scanned so intensively. By that time I was making another delayed discovery. To my astonishment—because he was older, had three children, and was only recently separated from his wife—I was beginning to understand what my heart already knew. I was falling in love with the rescuer who had sat facing me, so solicitously pantomiming warning. The rescuer's name was Doug Fesler.

※ BY 1:30—as my plane was nearing Cordova—Doug had pulled the linemen for whom he was responsible back into the safety of the tunnel. Visibility was down to less than a hundred feet, and pickup trucks were trapped in hood-high drifts. Other crews from various companies were working in the area, around forty men in all. Doug eventually managed to convince the lead contractor that if the whole project wasn't shut down soon, everyone might be spending the night in the tunnel, maybe several nights. Backhoes wrenched the stuck trucks free, and a procession of vehicles set out for Portage.

On board the Cessna Citation, the five of us didn't talk. It would have been hard to hear over the engine noise, and there was nothing comforting to say. Most of the hour-plus flight felt like a roller-coaster ride with intermittent free falls. As we descended, I caught glimpses of the mountains and, leaning forward, scrutinized them for clues that might help me in Cordova. How hard was it snowing? How much snow was the wind still scraping off the ridgetops? On what aspects was the loading occurring? How deep were the avalanches breaking? How widespread was the avalanche activity? The questions came easily, but the gaps in the clouds were too fleeting for answers. As we neared the town, I wedged my hips firmly into the seat and braced my feet on the floor, acutely aware that, given the crosswinds, we were likely approaching the runway sideways. I didn't even see the ground until I guessed we were ten feet above it, though the pilot later told me that visibility was a magnanimous hundred feet. The plane's tires bounced hard once, twice, and then again, until the rubber found a grip on the icy concrete.

We hurried, more out of habit than hope, into the dark empty terminal. It had been cut off from town by the avalanche, which had also wiped out the electricity and heat. Two startled Alaska Airlines agents scuttled out of the gloom to greet us. One of them immediately offered to drive us to 5.5 Mile.

❄ I AM ACCUSTOMED to being one of the first rescuers to arrive at an avalanche accident, usually after a hurried helicopter flight or a spine-rearranging snowmachine ride. I use those first minutes to gauge the lingering hazard and to interview any witnesses in an attempt to determine where the victims are most likely buried. Never before had I been driven to the edge of an avalanche in a warm car, unfastened my seat belt, and stepped directly into the bedlam of a search already under way.

Some of the sweat-soaked shovelers tackling the remains of Jerry and Martha's house had dug themselves into pits so deep that I could see only the tops of their heads. Safety seemed a ludicrous ivory-tower concept. The slippery, uneven debris was more hazardous than any I had ever seen, insidiously laced with wire, broken glass, wooden barbs, sharp metal edges, and leaking fuel. Lumbering amid the searchers were beeping bulldozers belching black smoke and orange Hitachi excavators with long swiveling arms grabbing overflowing bites of wreckage. None of the machines could take a pure bladeful of snow in the ghastly confusion of building materials and broken possessions. The destruction seemed total. Then, improbably, a whole teacup on an unchipped saucer, a clean sweater, or a compact disc cocooned in its case would be plucked from the chaos. I was struck by the intimacy of the task at hand; the rescuers were essentially quarrying through pieces of Jerry and Martha's private lives.

Dewey Whetsell had requested search dogs from Anchorage, believing them to be Jerry's only chance. Instead, Whetsell got me. He was, nevertheless, gracious in his welcome. Both he and Plumb hoped that I would know the answers to the questions gnawing at them. Was it safe to be searching for Jerry, and how long could they continue? Whetsell noticed that the searchers—exhausted by exertion, tedium, and a sense of futility—were getting that "glassy stare that people get when their brains go numb." Moreover, the storm hadn't abated. The slopes above were being reloaded, and as darkness encroached, the avalanche guard soon wouldn't be able to see well enough to provide even a yelp of warning. With each hour, the probability of a second avalanche increased while the probability of Jerry being alive plummeted. Whetsell and Plumb were unwilling to play with the lives of the fifty or more rescue workers on site, especially given the likelihood that this was no longer a rescue but a body recovery effort. And yet they knew with every nerve in their bodies that, if by the quirkiest of odds, Jerry LeMaster was still alive, abandoning the site for the night would mean pulling the plug on their good friend and coworker. They had just emerged from a huddle, where they had agreed to begin to scale back the number of diggers. By nightfall, no more than two hours away, the site would be emptied of all but a small core of searchers.

Even with almost two decades of experience evaluating avalanche hazard, I found it difficult to offer a second opinion. I didn't have X-ray vision and couldn't see any farther up the mountain than the other search leaders. The den of any lurking avalanche dragons was so completely veiled that it might as well have been the Forbidden City of Lhasa. I chose to make the same assumption, valid or not, as Whetsell and Plumb: given the mass of the avalanche, it had likely involved much of the available snow, buying an unknown number of hours before the slopes could be primed again. As a technical adviser with a bagful of rescue equipment, I had a few things I could do. I put avalanche beacons on the heavy equipment operators who were working deep in a dangerous trench to clear the Copper River Highway and assigned spotters to watch them.

Dewey Whetsell later confided, "I was *acting* like I had a plan, a system, like I knew what I was doing. In my entire career, I had never been so stumped by an operation. I was faking it completely, but I kept the demeanor." For lack of better options, he stuck with his original idea of quelling the din of saws, diesel engines, and radios every fifteen minutes so that dispatch could dial a tone-out while rescuers leaned close to the snow, straining to hear Jerry's fire department pager. Whenever a rafter or panel of plywood was extracted, leaving a larger-than-usual void behind, Whetsell instructed a digger to lower his head into the dark space and yell Jerry's name. I assured him that this plan was as good as any. When he asked what I thought of the chances of finding Jerry alive, though, I could only shake my head pessimistically.

By 3:00 P.M., five hours after the avalanche and an hour before dark, so many pager tests had been conducted that not even the most optimistic could muster hope of a response. We stood in small clusters, reacquainting ourselves with silence, when someone screamed, "He's here! He's down here! He just yelled!"

People sprang to life, wielding chain saws and shovels and heaving on timbers with newfound strength. Alarmed, I leaned into Bob Plumb's shoulder, cupped my hands, and shouted toward his ear, "We're going to kill him. We need to slow down!" For avalanche victims, rescue is often the most traumatic part of their ordeal, because shovelers are standing on them, trampling their precious airspace.

From a high hump of the debris, with radio in hand and the chin straps of his insulated blue hat flapping like a dog's ears, Plumb hoarsely stopped the action and demanded the diggers' attention. Gesturing to me, he said, "This is Jill Fredston. Probably the most expert person we've got at a point like this. She has good advice. Take her direction on this."

In the footage shot by the television crew I flew to Cordova with, my eyes look as intense as full moons. I knew that if we were going to get Jerry out alive, we needed to do three things. To rescue him quickly, we had to move slowly. Equally counterintuitively, we couldn't dig straight down to him. Instead, we needed to shovel our way in from the side to avoid collapsing whatever airspace he had or inadvertently dropping a piece of wreckage onto him. Finally, we needed to stop banging heads and frenzied shovels, and work deliberately as a team. Jerry had probably been buried fifteen feet under the snow surface, but because rescuers had already removed tons of debris, he was now about six feet below us. Extricating someone from that depth requires a hole at least twice as wide, with the diggers closest to the center throwing shovelfuls of debris onto an intermediate shelf, more diggers moving that tier toward the top of the hole, and still more diggers keeping the rim clear. But the situation was complicated by large pieces of plywood, rafters, and metal extending into the walls of our pit mine. Often we couldn't use the saws until we cleared the area around the obstacle and made sure that its removal wouldn't catapult something else onto Jerry. I say "we," but I rarely plied a shovel or saw. I've learned that I can best lead an avalanche rescue by doing as little physical labor as possible. Instead, I stand back: watching, delegating, anticipating. As a result, I'm easily spotted at a rescue site. I'm the one wearing the most clothes.

The footage aired on television made it look as though we freed Jerry in minutes. In reality, it took a quarter of an hour just to reach a small hole in the snow through which, with the help of a flashlight, we could see the top of Jerry's head. Volunteer firefighter Mark Kirko crawled over, stuck his face into the hole, and wriggled out to report that Jerry was conscious, but very cold and unable to move one arm. A woman medic who could squeeze her shoulders into the hole traded places with Mark. Once there she fed an oxygen tube into the void to supplement Jerry's air supply and placed heat packs near the top of his head, which was as far as she could reach. Jerry wasn't laid out nice and neat as if in a bed—or, more realistically, a coffin—which made it difficult to know where we could safely dig. He was lying on his side, trapped in a triangular airspace formed when the furnace and hot water heater landed slumped against each other like a couple of drunks. Another five minutes of digging went by, then ten, then twenty. Joanie Behrends, the medic, kept reporting that Jerry felt increasing pressure against his chest. What were we moving that was making it harder and harder for him to breathe? With a rope and the collective force of eager arms, we were able to tug the water heater out of the hole. The furnace was more problematic because it was wedged against the same blue door that Jerry had hold of when the avalanche hit. Now the door seemed to be pinning Jerry's midriff, suffocating him. A few men wrapped a rope around a troublesome timber and used a backhoe to wrench it free, but we were afraid to use the equipment to pull out the blue door because we thought, depending upon how Jerry was entangled, that we might rip him apart.

The diggers had been working ferociously for an hour when Joanie jerked her head out of the hole: Jerry was unconscious. We no longer had options. With a nod from Whetsell, one man grabbed a chain saw, yanked the starting cord, and sliced the blue door lengthwise—though none of us had any idea whether he might be amputating one of Jerry's limbs. Joanie, still pressed as close to Jerry as she could get, barked, "He's not breathing!" Medical protocol cast aside, numerous hands reached forward to pull Jerry from the hole.

I knew as soon as I saw Jerry that we hadn't moved the blue door fast enough. His face was the dull purple-gray color of the sky and ominously bloated. His friends and neighbors tried to cheer him back to life, shouting, "Jerry!" "Breathe, buddy!" "Come on, Jerry!" A chorus of *If only we had*s began to sound inside my head, even as we threw Jerry on a stretcher and, without taking the time to strap him down, ran across the debris to a waiting ambulance. Tears streamed down faces unaccustomed to crying as we watched the receding red lights signal an emergency that was no longer ours. My watch read 4:04 P.M. Jerry had been buried for just over six hours.

✳ AT 4:00 P.M. Doug and the tunnel project workers reached Portage, where they found twenty-six inches of new snow on the ground. They crawled the five miles back to the

Seward Highway in a convoy, rarely moving faster than ten miles per hour. Visibility was so poor and the new snow so thick that Doug—in the lead, with chains fastened on all four wheels of our Suburban—sometimes had to stop and walk in front of the headlights to find the road with his feet. He didn't yet suspect that it would be several weeks before a barrage of storms released its grip on southcentral Alaska and allowed workers back into the tunnel.

Doug first heard of the 5.5 Mile avalanche from a construction worker. A few news snippets filtered through the static of the car radio and the steady plod of the windshield wipers on the drive back to Anchorage. He knew that the Troopers would have called me to assist but was incredulous when he discovered my car gone from the driveway and the house dark. "I was jazzed by the storm when I reached home," he told me, "worried when I read your note in the kitchen, and furious by the time I reached the upstairs phone to call the Troopers for an update. It was the third worst blizzard I'd ever been in and the absolute worst at sea level. I couldn't believe you had taken the risk of flying in that crap."

❄ WHAT KILLED JERRY was probably hypothermia along with "compartment syndrome," which used to be called "tourniquet shock." Denied oxygen, the cells in his crushed arm began to burn glucose in a last-ditch effort to survive. A toxic by-product of this process is lactic acid, which built to lethal concentrations during the long hours of burial. When we freed Jerry, the lactic acid in his blood flooded back into circulation, stopping his heart.

What saved Jerry were the efforts of his medic friends, who pounded on his bruised chest and pumped air into his cold lungs as the ambulance sped toward town. By the time they reached the hospital, they had restored Jerry's pulse and he was talking, albeit weakly. When dispatch radioed that Jerry was alive, a cheer erupted among the troops dejectedly gathering gear in the dark at 5.5 Mile and, as though a switch had been flipped, we began to banter.

Jerry later recalled that the first time he yelled was when he realized something terrible might have happened to Martha—three seconds into the ordeal, while his mind was bouncing everywhere. Because his chest and back were compressed by unyielding objects, he could only manage 20 or 30 percent of an ordinary breath, so even feeble cries emptied him of air. Nor did he have a way to measure time. He had never felt the need for a watch, and even if he'd been wearing one, he wouldn't have been able to see it. As he struggled to maintain consciousness and shivered uncontrollably, time slowed. He thought that the boards he was kicking were the floor of the room above and that his rescuers were only inches away, when, in fact, we were floundering fifteen feet above him, out of earshot. Occasionally he would yell, "Where the hell are you?" After a while he began to slip in and out of consciousness, and no longer worried about his fate. Finally, he could hear saws. When they quit, he would yell, but when the saws started up again, they sounded even farther away, and "that was not a good feeling at all." Perhaps because of settling of the snow debris, the weight squeezing his chest grew progressively heavier. "Quite honestly," Jerry said, the yell we finally heard was his "last act of desperation."

When Mark Kirko pushed his head into the hole and asked, "That you, Jerry?" Jerry thought, *Who the hell did you think it would be?*—but didn't have the breath to say it. He became more and more anxious for us to remove the elephant from his chest and couldn't understand our ineptitude. As we had feared, we shifted something toward the end, perhaps the furnace or the recalcitrant blue door. Whatever it was most likely moved less than an inch, but that was enough. Jerry lapsed into unconsciousness, knowing that his chest was gripped in a vise so tight that he could no longer breathe on his own.

❄ THREE SMALL PLANES left Cordova that night. The first carried the search dogs and their handlers, who had landed just as word came that they weren't needed. Literally carrying the news, ours was the second. Turning around behind us, almost as soon as it arrived was a specially equipped medevac plane hurrying Jerry to the hospital in Anchorage. Though it still wasn't a night for flying, a lull in the storm combined with an emotional rush akin to winning the jackpot made the flight far shorter and softened the terror.

I reached home in time to catch footage of Jerry's rescue on the ten o'clock news, though my eyes kept blurring beyond focus as I tried to watch. Doug set aside his anger so we could lie together in quiet celebration with our scruffy black dog, Bodie, belly up on the bed between us, his legs raised like goalposts. Avalanches were integral to our courtship and have remained a third vociferous partner in our marriage. On nights we've unburied the dead, their faces slip into bed with us, interrupting sleep with a relentless replay of the bodies we've hauled out of the mountains. After decades of avalanche death, this one night was a reprieve.

After the headline news, we kept the television volume muted until it was time for the weather forecast. In truth, we didn't need to hear the weatherman's spiel, for the satellite imagery illustrated what we already knew. Alaska was under assault by a fire hose of subtropical moisture-laden air streaming from the South Pacific. On the weather map, parallel lines of equal pressure over the southern mainland corseted so closely together that they looked like a cliff on a topographic map. This tight pressure gradient and steady wet flow, dubbed the "Pineapple Express" by National Weather Service forecasters, meant that we were only a couple of battles into a storm that was far from over.

CHAPTER 2

Union of Circumstance

The author investigating a fracture line in the Kenai Mountains
© DOUG FESLER

*Maybe falling in love, the piercing knowledge that we ourselves
will someday die, and the love of snow are in reality not
some sudden events, maybe they are always present.*
—Peter Høeg, *Smilla's Sense of Snow*

*The winter . . . is thrown to us like a bone to a famishing dog,
and we are expected to get the marrow out of it.*
—Henry Thoreau, American naturalist and writer

WHEN DOUG THUMBED INTO ALASKA AS A FRESH-FACED
college student on break in the summer of 1966, his avalanche knowledge consisted of the little he could remember
from an atrociously spelled two-page book report he'd prepared while struggling through the seventh grade. Now he
was majoring in sociology and education, disciplines that
would prove surprisingly useful when he eventually turned
his attention to avalanches.

Lured by Alaska's wildness, Doug had hitchhiked from
North Dakota. At the border in Sweetgrass, Montana, Doug
fell $135 short of the $150 he had to display to gain entry to
Canada. His brother wired him the funds, and from the first
town he reached in Alaska, Doug wired them back. He was
left with $7.28 in his pocket, a homemade backpack welded
from steel conduit, and a back strengthened by days spent
lifting rocks and throwing hay bales on North Dakota farms.
"It took a week of hard traveling along a winding dusty ribbon of a road mounded with frost heaves just to reach
Alaska," says Doug. "I didn't know anyone, but even on my
first day, I knew I was home." Unable to afford the YMCA's

$2-per-night rate, Doug bunked next door at the Anchorage Rescue Mission, a church that had been converted into a homeless shelter. In exchange for performing maintenance chores, he also received "lots of Jesus," a hot supper, and breakfast. "Fortunately," he says with a grin, "I like singing."

Work was scarce in pre-pipeline Alaska, especially for outsiders, so Doug scavenged odd jobs. He hauled construction supplies, dug ditches, ran a road-striping machine, shoveled gravel, jackhammered guard rails, spiked railroad ties, helped erect a Ferris wheel, and painted houses. He did, though, turn down a grunt job in a remote gold mine after he heard the boss was slow to pay wages and fed his captive workers only pancakes.

Alaska was bigger and rawer than any place he'd ever been, with more room to explore and more freedom to make choices in a place "where everything didn't already feel all figured out." Even deep in the New England woods, where he spent much of his childhood, he'd come upon stone walls that had once marked the edges of fields cleared by colonists in the 1600s. Doug had left home as a teenager, eventually making his way to North Dakota. An aunt had convinced him to finish high school there, which he did while working forty-six hours a week, and he had gone on to enroll at North Dakota State University. His grandmother Josie, born in 1881, had been among the last of the homesteaders in Dakota Territory. Homesteading was already history in the West, but Alaska was still a frontier, where the hardy could stake claim to a piece of land. As Doug journeyed north, he'd seen tidy ranches on open prairie give way to rough cabins set on the edges of newly cleared fields, which in turn yielded to uncut expanses of trees. Alaska had more of everything Doug already loved— more mountains, more wildlife, more untracked places.

After another year of school, Doug returned to Anchorage for a second summer. This time he worked as an aide in a psychiatric hospital, taking inmates on outings. While driving the patients south to view the lake and glacier at Portage, he was so awed by the great swaths of forest uprooted and minced by avalanches that he slowed the bus to a stop, trying to fathom the forces and imagine Alaska in winter.

Doug graduated from college and moved north for good in 1969. He reasoned that North Dakota, with blizzards that buried barns to their eaves and bitter winds, had been good training ground for Alaska winters. In a memoir titled *Dog Puncher on the Yukon,* he'd read of a trader during the mad rush for gold at the turn of the century who had fashioned a special thermometer to gauge the lung-burning cold of interior Alaska. It consisted of "a set of vials pushed into a rack, one containing quicksilver, one the best whiskey in the country, one kerosene, and one Perry Davis's Pain-Killer. These congealed in the order mentioned, and a man starting on a journey started with a smile at frozen quicksilver, still went at whiskey, hesitated at the kerosene, and dived back into his cabin when the Pain-Killer lay down." But to his surprise, Doug found that with temperatures moderated by the ocean, Anchorage's climate was generally more benign than that of North Dakota. On the shortest day of the year at such latitude, though, the sun managed to rise only a feeble six degrees above the horizon. During the rare stretches when the temperature refused to budge above −40 degrees Fahrenheit, unfazed, Doug "just put on more sweaters."

His first full-time job in Alaska was behind bars in a juvenile correctional facility, where he became the recreation coordinator. Until then the sole sources of recreation had been a scuffed pool table and a television set for each lockup unit. Seeing more therapeutic value in the mountains, Doug began taking his charges on three-day hiking trips that soon became coveted privileges. Most of his spare hours were also spent roaming the mountains, where he deepened his acquaintance with snow.

Doug's memory of snow dates to when he was a four-year-old in rural Illinois. He marvels at the freedom he was given; his mother would routinely bundle him into a brown canvas snowsuit, pull down the earflaps on his wool "bomber" hat, and send him out the door to play for hours at a time. Across the street was a cornfield covered with three feet of snow. A prolonged cold spell had given this snow a loose granular texture, while a more recent warm storm had capped it with a harder two-inch-thick ice crust. Of course, as a four-year-old, Doug was aware only that he could tunnel on his belly through the almost resistance-less snow and pop, like a rabbit, through the crust anywhere in the vast field. He could even scoot along the surface of the crust, something his brother and his brother's friends—who were more than four years older and thus heavier—could not do. Doug started downhill skiing that same year, gamely balanced on the back of his mother's skis, his short arms wrapped around her legs.

Doug couldn't get enough of winter in Alaska. On his forays into the hills, he rarely encountered people, though he thrilled to find signs of plentiful wildlife. In April 1970, when I was a twelve-year-old tomboy dutifully attending junior high school in suburban New York, Doug was on cross-country skis in the Chugach Mountains thirty-five miles southeast of Anchorage, following a friend across a broad treeless slope that dropped into a gorge three hundred yards below him. In summer Crow Creek churns through this ravine and a well-trodden trail switchbacks uphill, but there was no trace of the trail late in the winter, with fourteen inches of new powder and the sky still bucketing snow. Doug and his partner were close enough to Alyeska Ski Resort that they could hear the cannon being shot to bring down avalanches. Doug had nearly reached dead center of the large slope when he felt and heard a *whumph* that seemed to begin beneath the soles of his boots and reverberate throughout the valley.

There was much Doug didn't know. He didn't know that the hippie friend in front of him would become a straitlaced fire chief and die ten years later in a high-impact car crash. He didn't know that thirty years later he would pull the body of a young woman just killed in a small avalanche from the same rocky gorge that gaped below him. He didn't even know that the *whumph* was the sound of a buried weak layer collapsing under the weight of the new snow or his body. But he knew that the *whumph* meant trouble and was likely the slope's final sigh before it avalanched. He also knew that if the avalanche carried him into the gorge, he would die. Doug yelled to his friend, who had reached the far side, to watch him. Then he willed his body immobile. Desperate not to disturb the slope, he didn't move for several protracted minutes. Then, as though "walking on eggshells," he began to trace his friend's tracks, creeping toward the safety of bare ground.

Doug had other encounters with unstable snow that spring, including a wet slide that cartwheeled him a thousand feet, burying him thoroughly enough that it took forty minutes to claw himself free. Still, he probably would not have made avalanches his passion if he hadn't been hired, in April 1971, as one of the first three rangers in Chugach State Park. The new park was the first of its kind in Alaska; no one had even located its boundaries on the ground when Doug began working. Infuriated that the government had "locked up their backyard," some of the homesteaders with neighboring property threatened to shoot any ranger who set foot on their land. The three rangers wore khaki shirts with Alaska State Parks patches sewn by hand onto their sleeves. They looked like gas station attendants, and the job was about as glamorous. Doug estimates that he drove a thousand miles a week emptying garbage cans and hauling firewood in the campgrounds and rest areas rimming the park.

Anchorage—Alaska's biggest city—is wedged between the Chugach Mountains and the gritty waters of a 200-mile-long ocean inlet named for British navigator James Cook. With peaks rising from sea level to over thirteen thousand feet, the Chugach is Alaska's fifth-highest range, not a minor claim in a state with thirty-nine distinct mountain ranges and seventeen of the twenty tallest mountains in the United States. Since the park, which encompasses a half-million acres of wilderness, was established in 1970, the population of Anchorage has doubled to around 275,000 and the city has become home to at least half the state's population. Yet it is still possible after driving twenty minutes from town to scale the first set of ridges barring the city's sprawl from view and hike or ski more than a hundred miles without reaching a road. It is also possible, as has been proven repeatedly, to die in an avalanche within sight of the city's shopping malls.

In the 1950s and 1960s, an average of only five people a year died in avalanches in the United States. In the 1970s, as winter recreation grew increasingly popular, the numbers began to soar, with twenty-four fatalities in Alaska alone. On Doug's very first day on the job, two experienced ski mountaineers were killed in an avalanche in Chugach State Park.

One of the victims was a physician who had lost her husband to an avalanche in the Himalayas only a few years before. Women die in avalanches far less frequently than men—a statistic that cannot be explained entirely by the greater number of men exposed. In *The Book of Risks*, Larry Laudan notes that there is never an age, from infancy through old age, when the male death rate does not far outpace that of females, and males have three times as many accidents. "If you want to stay alive in the mountains," Doug now says, only partially joking, "travel with a good woman and listen to her." On the day Dr. Grace Hoeman was killed, Doug and the other rookie rangers didn't even venture to the accident site, though it was within their jurisdiction. They had to defer the multiday rescue effort to volunteers with the proficiency to know how and where to search.

Doug's avalanche rescue experience mounted quickly. During one search, he helped trail drops of blood and strands of hair downhill to find the body of a sixteen-year-old climber buried under only eighteen inches of snow. Doug was on his knees digging at another accident site when he uncovered a frizz of hair standing on end as though jolted by an electric shock. The openmouthed horror-stricken face of a twenty-four-year-old librarian began to take shape under his hands. It is a face Doug remembers in painful detail forty years later.

During the 1970s, before frontier attitudes softened, Alaskan avalanche victims stayed buried for an average of 140 days—a statistic that does not include those who were never found. Searches were suspended when they had clearly become recovery efforts, and bodies were left to melt out in spring. With time as the enemy, a buried victim has the greatest hope of survival if his partners stay on-site and search rather than leaving to seek help. Summoned rescuers have about as much chance of changing the final score as third-string basketball players sent in to play the final minutes of a game that their team is badly losing.

By 1973 Doug's garbage detail had given way to search and rescue, law enforcement, and interpretive duties, and Doug was patrolling the park regularly, on foot, skis, and snowmachine. He had also become an active member of the Alaska Mountain Rescue Group, a volunteer squad on call statewide. His enthusiasm was akin to that of the archetypal fireman who sleeps with his boots on and flings himself down the fire pole at the first alarm bell. At any hour and in any weather, he might be summoned to climb to a plane that had crashed onto a glacier riddled with crevasses, to lower an injured goat hunter off the thin ledge of a cliff, or to find a child lost in the woods. Doug likens rescue missions to chess games for the careful unfolding of skill and strategy. And from the camaraderie of shared challenges and close calls arose friendships that transcended beer and pizza and have stood solidly through the years like pillars of erosion-resistant rock.

As 1974 rolled in, two more avalanche fatalities occurred in Chugach State Park within two days of each other and scant miles apart. "Those two events caught my attention big-time," Doug says. "Both places where these guys were killed were places I wouldn't have thought twice about crossing in winter. I realized that I really needed to learn something about what was going on here, or I was going to wind up getting killed. Because I was going out all the time, traveling in that terrain; I mean *all* the time."

Since rescue clearly wasn't the answer, Doug turned his energy toward the questions. Where did avalanches happen? When was the threat greatest? What made a slope unstable?

Could avalanches be forecast? Was it possible to travel safely through the mountains in winter? Could he teach himself and others how to avoid becoming avalanche victims?

Alaska was relatively isolated in the pre-Internet dark ages of the 1970s, removed even from the growing avalanche community in the Lower 48 and Canada. "If anybody had any avalanche-related publication," Doug remembers, "I would borrow it, copy it, read it, file it, go back to it, study it, and think about it." He invested at least two hours a night chipping away at the literature stacked beside his bed. Much of the information he gleaned felt like "a shotgun blast to the face. I spent most of my time trying to figure out which of the pellets were more important than the others."

When Doug read that snow crystals change constantly throughout the winter, he thought, *Aha! Snow metamorphism is the secret I've been looking for.* Snowflakes are deposited or rearranged by wind not into a single "blanket," as the winter snowpack is often poetically described, but into distinct sheets of varying thickness and hardness. The changes that then take place influence how well individual grains of snow bond to one another both within a layer and between layers. That some layers become strong while others progressively weaken is what sets the stage for avalanches.

Day after day Doug climbed into the mountains to dig holes in the snow, sometimes burrowing twelve feet to the ground. Like a cold-adapted Sherlock Holmes, he'd poke his fingers into the pit walls and make even parallel strokes with a paintbrush to identify strong ridged layers and weaker indented layers, and then examine the snow crystals under a high-powered hand lens. Doug's dyslexia had made school a torment, but by the time he discovered that snow metamorphism was a piece of the puzzle rather than the solution, he had grown skilled at reading the stories embedded in the snow. The more he learned, the stronger the pull of his curiosity.

Even as his learning curve was climbing with the trajectory of a jet on takeoff, Doug felt a pressing responsibility to help stem the surge of backcountry avalanche accidents by passing on his new knowledge. He founded the Alaska Avalanche School in 1977 and began offering hands-on workshops that focused on teaching participants how to move through snowy mountains safely. He and other rangers began issuing weather and avalanche advisories for those venturing into Chugach State Park. This grassroots forecasting effort evolved into the Alaska Avalanche Forecast Center, which produced advisories for three separate mountain ranges in southcentral Alaska by the time I arrived in Anchorage in the summer of 1982.

❄ CLUTCHING A brand-new master's degree in polar studies and ice from the University of Cambridge in England, I had hurried to Alaska—one of the few places I might actually find a job. Almost immediately I stumbled into a position as a snow and ice specialist for the University of Alaska. Anything frozen was deemed my purview—I worked on tidewater glaciers, solid seas, and ice-jammed rivers. Within a few months of my arrival, the university inherited the Alaska Avalanche Forecast Center, and it was proposed that since I knew something about snow, I be named the director and lead forecaster. There was only one problem: I knew nothing about avalanches. My ignorance didn't faze any of the overseers except for the state's reigning avalanche authority—Doug Fesler. By then Doug's expertise was so widely recognized that his name was drawn into nearly any conversation about ava-

lanches in Alaska. Though Doug was running a separate avalanche safety program for a different government agency
and was not responsible for the forecast center, my résumé
was sent to him for review. The forecast center had slipped
under the rule of meteorologists, one of whom couldn't ski
and knew more about the weather on Mars (the subject of
his master's thesis) than he did about avalanches. Fed up
with seeing the credibility of the forecast center compro-
mised by unqualified forecasters and unimpressed by the
"fancy schools" on my résumé, Doug urged against my ap-
pointment. He was right—but I was handed the job.

I traipsed across town to introduce myself to Doug,
blithely unaware that he thought me "as green as they come."
Thirteen years my senior, Doug seemed to know everything.
His subterranean office was intimidating, a tangle of climbing ropes and hard-used skis leaning against walls pinned
with photographs of incomprehensibly huge slides and the
twisted cars and roofless houses they'd left in their wake. He
remembers me as a twenty-four-year-old who had never
even seen an avalanche, perched on a stool with a sharp pencil poised over a blank notebook.

I wasn't entirely without credentials. I'd been fascinated
by snow since I was a six-year-old with a plastic shovel in the
backyard—a somewhat inexplicable obsession since annual
snowfall was thin where I grew up, in a suburb north of New
York City. My exposure to more serious snow was limited to
family ski vacations. By the end of college, though, I'd found
a field assistant's job on a surging glacier in Canada, and in
preparation for my master's thesis, I spent months in Green-
land scrutinizing a core drilled more than a mile deep into
the ice sheet. The chronological layers embedded in the core
could, as science writer Walter Sullivan wrote in the *New*

York Times in 1981, "reveal the story of the snows that fell when Cro-Magnon artists were painting the images of prehistoric animals on the walls of French caves." In the deeper reaches of the core, ten thousand years of ice were compressed into less than a vertical foot, rendering time such a blink that one millennium was not always distinguishable from the next.

My master's program seemed esoteric to most, including my ever-generous parents, who helped foot the bill, but it took me exactly where I wanted to be—to a thinking job outdoors. I wasn't interested in science so much as in applying the science: avalanche forecasting felt like a logical and good fit. Only now am I willing to admit that it was a rather abstract choice. Like James Joyce's character Mr. Duffy, whom he described as living "a short distance from his body," I was buoyed by an equally abrupt—and possibly dangerous—disconnect between my youthful confidence and my qualifications.

Alaska offered what I craved. I'd tasted five weeks of Alaska's wildness in a sea kayak when I was sixteen and was champing at the bit for more. I wanted mountains and meadows and sea to be a part of my life rather than vacation destinations. I hoped moose and bears would wander freely through my yard. I wanted to be defined as much by what I loved as by what I did for a living or where I'd gone to school. Alaska, with its remoteness and breadth, seemed to afford greater room to live, to be, to wander. And so, close as I was to my family, I moved across the continent.

Much as it had for Doug, coming to Alaska gave me the feeling that I'd left the barriers of convention behind, that everything was possible. If I wanted to row the coast, I need only get a boat. If I wanted to build a house—though I'd never even flipped the switch on a power saw—well, then, I'd better get going. And if I was going to forecast avalanches, a good place to start was to pick the brain of Doug Fesler.

Doug made it clear that if I was to have any chance of bringing avalanches into focus, I needed to learn to read the history of a single winter's weather in a snowpit wall. If terrain is the foundation of avalanches and weather the chief architect, then the snowpack is the blueprint. Doug encouraged me to translate book knowledge about snow metamorphism into understanding by setting up a study plot in my backyard. After each storm, I laid colored string on top of the snow between two trees. Every few days I dug pits and observed changes in the layers. That winter I watched eight inches of fluffy powder compact to four inches of ever-smaller, stronger, more rounded grains under cloudy warm conditions and witnessed clear windless weather produce surface hoar, fragile feathers of frost that are capable of wreaking havoc months after they have been buried by subsequent storms. During long cold, dry spells, when many might believe the snowpack dormant, I monitored the growth of large grains with the flashy facets of diamonds, keenly aware that this classically weak layer causes more than its fair share of mayhem and heartbreak.

But in our first meeting, leaning back in his chair, prominently veined hands knitted behind his head, slate eyes squinting, his expression inscrutable behind a bushy brown beard that lent him the appearance of Moses, Doug cautioned me not to spend all my time in my backyard. "If you really want to learn where the avalanche dragons live and feed, when they sleep, and what fires them to life," he said, as nonchalantly as if he were sending me to the corner grocery for a quart of milk, "you have to hang out in the den of the dragons."

Though I had long loved snow, I disliked cold. A string bean of a kid, I took every opportunity to whine when I anticipated discomfort. I'd been enticed north by the prospect of bigger and bigger mountains, but I was so leery of heights that I had flunked slides in kindergarten. I enjoyed skiing and I was a decent downhill skier, but it was not a calling. I had no experience climbing uphill on skinny telemark skis that allowed my heels to flap freely or descending similarly untethered through challengingly variable snow. I had, however, been a diligent student, and Doug's assignment didn't sound optional.

I began to haunt the mountains, frequently alone because I couldn't recruit partners to scale ridges in storm winds that were sometimes so strong they could fold my knees like a supplicant's, reducing me to a crawl. Swaddled in nearly every piece of warm clothing I owned, I'd tie one end of a climbing rope to a big rock with as many knots as I could muster, hope for the best, and lower myself into the lair.

❄ EARLY INHABITANTS of the Alps regarded avalanches not as comprehensible natural phenomena but as monsters that might be kept at bay by ringing the village church bells with vigor. Mountain folklore from Europe is rich with references to avalanches as living creatures; translated from German, one such example is, "What flies without wings, strikes without hand and sees without eyes—the avalanche beast." As late as 1652, a trial in Switzerland affirmed that "witches are the causes of avalanches." The dragon in my sights was called a slab avalanche, one of four species of avalanches, but the type responsible for most avalanche accidents worldwide.

Though the word *slab* conjures images of something solid like a side of beef, slab snow is simply one or more layers of snow that are more cohesive than the layer below. Picture an eight-year-old boy proudly carrying his father's double-decker birthday cake to the table when, in the dim light and excitement, he tips the cake at too steep an angle. Depending on how the cake is constructed, it might bear the additional stress without catastrophe. Maybe only the chocolate frosting on top slides free. Maybe the upper half of the cake fails on the relatively weak middle layer of jelly. Maybe the whole cake avalanches off the plate, which acts as a bed or sliding surface.

Some slabs are made of snow so improbably hard and strong that they could pinch-hit for cement, but strong is far from synonymous with stable when weaker snow is underneath. Others are made of seductively soft snow that can't be picked up without crumbling like sawdust. Slabs can be hard or soft, wet or dry, big or small. It is the slab's cohesiveness that is important because it is this property that makes the snow capable of breaking rapidly on a broad front. The slab tears loose from a slope as one large block before rupturing into smaller pieces.

Just as a bowl filled with flour, sugar, and eggs does not necessarily make a cake, there is more to a slab avalanche than the presence of a slab, a weak layer, and a bed surface. For the recipe to be complete, the strength of the snow must be nearly equal to the stress being exerted upon it—and something must tip the balance. Storms are common triggers, as are people.

Imagine that you have met the strongest man in the world. He is predictably huge, with a neck the size of your thigh and hands that look like catcher's mitts. You are charged with the seemingly impossible task of bringing this giant to his knees. You lie awake fretting and scheming night after night until finally you hit upon two plans. The first is to load a mere 100-pound sack onto his vast shoulders, and then leave him without food, water, or encouragement. You are confident that, given enough time, he will weaken and crumple to the floor in a bulky heap. The second plan is to stockpile 100-pound sacks and heave them onto the strong man's shoulders as fast as you can, giving him no time to adjust to the added weight. He might be capable of withstanding colossal stress, but you are betting that when the load finally draws even with his strength, he will collapse, and you will prevail. As with people, some snowpacks are stronger and less susceptible to stress than others. But no snowpack can tolerate an infinite amount of stress—and the faster stress is applied, the sooner the snow reaches its breaking point.

The mechanics of slab fracture are hard to study without the expensive sacrifice of precariously perched instruments, but measurements indicate that the bonds holding a slab in place can fracture at speeds greater than two hundred miles per hour, which is why a mile-wide slab is able to release virtually instantaneously. This also explains why triggering a slab feels like having a carpet yanked from underfoot, a sensation I discovered with an equal surge of gratification and adrenaline after a few weeks of eager experimentation on short steep hills and road banks.

✳ SNOW VOICES complaint in a variety of ways. It is unsubtly screaming distress when it makes *whumphing* noises or sends long, thin cracks arcing across a slope. But during my early days in the dragon's den, the unhappier the snow, the happier and more intrigued I became. Nothing thrilled me more than finding a small slope so tender that one jump with my 125-pound body could make thousands of pounds of snow fracture like plate glass and careen downhill. No sooner had the debris come to rest at the bottom than I'd be scrambling over the fracture line—the perpendicular wall of snow left behind at the origin of an avalanche—to prod the layers and measure the steepness of the slope. Almost as exhilarating as the avalanches themselves was the experience of learning a new line of questioning. Making discoveries through observation, I sometimes sat in the mountains for hours monitoring the rate at which new snow accumulated, the temperature rose, or the wind blew. Watching the slope for signs of instability was like keeping vigil at a deathbed, except that I was waiting for the snow to come alive.

When Doug first preached the importance of developing "avalanche eyeballs," I'd nodded agreeably but privately thought him a madman. Quickly, though, I began seeing things I'd never noticed before. Thin wavy filaments clinging to the sides of telephone poles were a graphic-enough demonstration of the elastic deformation of snow to make me stop my car and snap photographs from the middle of a crowded sidewalk. Snow curving gracefully from the overhangs of roofs during warm weather made me wonder what effect the rising temperature was having on the stability of the snowpack in the mountains. I couldn't shovel new snow off my walkway without first measuring its density. I began to see slab avalanches everywhere. My thoughts always seemed folded in among the layers of the snowpack, and I slept with an ear tuned to the weather. Though one field book now sees me through a winter, in my first season I filled more than eight thick books with crabbed notes. One holds my observations of a slab with a fracture line an inch thick and eighteen inches long, with flanks of eleven inches. It ran

Snow is a unique material; it can flow like honey or fracture like glass. © DOUG FESLER

on a glassy bed surface that had a slope angle of 49 degrees. This particular specimen let loose on the windshield of my car. Nowadays I can't land at a snowy airport anywhere—including avalanche-free Chicago—without noting the wind direction from the sculpted patterns on the snow.

Much is made of the expanded lexicon indigenous people of the Arctic have developed to describe the states and moods of snow. "Dozens of words," people from snowless places exclaim, "just to describe snow!" To me, the surprise is that so few people realize that we have as many terms in English, among them *powder, stellar crystals, spatial dendrites, graupel, corn snow, hoar frost,* and *wild snow.* The environments in which we find ourselves shape our personal vocabularies. I couldn't be more thankful that I haven't had to employ the same nuanced vocabulary for traffic jams exercised by those who make daily commutes to crowded cities.

❄ A TYPICAL FORECAST morning started long before any part of me wanted to be awake. Prying myself from bed by 3:30 involved keeping the bedtime of a preschooler and staging three alarm clocks between the edge of the bed and the bottom of the shower. Having never acquired a taste for coffee, I relied upon an economy of movement to dress and get myself out the door. On cold mornings, by the time I reached my car, the ends of my wet hair were frozen and clinking like wind chimes. The forecast center occupied a cubicle provided by the National Weather Service in the Federal Building downtown, and as I shuffled to my desk, the meteorologists who had been on shift all night would look up from their glowing screens and ask what the weather was doing outside. By 4:30 I was propped in front of my own terminal, trying to divine what weather lay in store for various mountain ranges and to anticipate how the snowpack was likely to respond. I sent the first forecast out before 7:00 to aid workers responsible for keeping transportation corridors and ski areas safe. An hour later I'd record weather and avalanche advisories onto a hotline for the recreational public who might just then be rubbing the sleep from their eyes and plotting a day of fun in the mountains.

I had inherited a forecast center with a bad reputation earned in part because my predecessors had been meteorologists who had relied entirely upon weather input to anticipate avalanche conditions and rarely ventured onto the slopes to see what was actually happening in the snow. Anxious to boost the accuracy of the forecasts, I tried to spend at least a few hours of each day in the field. As I clomped out of the office in my mountain gear, my boss would ask, rolling his eyes, "Going skiing, *again*?"

During my rookie season, Doug called the forecast center one March morning after I'd issued the forecasts and casually asked if I'd like to helicopter with him and a partner to a site near Portage to investigate a recent avalanche that was rumored to be enormous. Like a little kid invited to tag along with the cool big brothers I never had, I hummed through the rest of my chores. I'd been in the field with Doug maybe five or ten times that winter, not often enough to shake—or take for granted—the excitement a benchwarmer feels when finally called into a game. I felt comfortable enough to approach Doug with questions, as long as I was reasonably sure they weren't idiotic. Doug always answered them patiently, clearly, and in detail. We were friendly, but we didn't go out of our way to be friends. I knew Doug was reading my advisories, biding his time, waiting to see if I was going to be another here-and-gone forecaster or, worse, a witless one.

Unlike tornadoes, avalanches do not touch down randomly but occur in fixed locations known as avalanche paths. The chief determinant of whether a slope can avalanche is its angle; for cold, dry snow, the tipping point starts at about 25 degrees. In case you are unaccustomed to keeping an inclinometer handy, the incline of an average staircase is 30 degrees. Slab avalanches are most common on slopes with angles between 35 and 45 degrees, the steepness of expert runs at ski resorts. Most avalanche paths with a history of threatening people or property earn names for themselves. Valerie's Climax near Snowbird Ski Resort in Utah acquired its moniker when Valerie and the man with whom she was having an affair defied orders to evacuate employee housing while avalanches were being shot down. After an avalanche ran through Valerie's house, the lovers had to be rescued naked from her crushed bedroom.

The path we were aiming for bore the less provocative name of Williwaw, an Aleut word meaning "strong winds," but it had a reputation for producing outlandishly large avalanches. In January 1980 rescuers ate the avalanche victim they found on the valley floor. The victim—a female moose— was still alive when highway workers helped extricate her, but days later, when she showed no signs of recovering from the ordeal, they "put her into a freezer of a different kind."

An avalanche path is the sum of three parts: the starting zone at the top, the track down which the avalanche accelerates, and the runout or deposition zone where the avalanche stops. As we clattered toward Portage, Doug directed the pilot to land in a summer campground more than a half mile from the mountain's base, which was still well within the Williwaw Path's runout zone. The avalanche had clearly made a forced entry. Where there should have been a two-seater outhouse with brown walls and a green roof, white toilets sat exposed on a concrete pad. Nothing much bigger than a matchstick was left of the outhouse walls. In the surrounding forest, trees the diameter of basketballs had been yanked from the frozen earth like weeds, and pulverized limbs littered the ground. Some standing trees had pebbles embedded like shrapnel in their trunks, forty feet above the ground. Picnic tables weighing 350 pounds had been tossed like Frisbees half the length of a football field.

The culprit was powder blast, the billowing leading edge of a fast-moving dry avalanche so singularly spectacular that it wins the starring role in any avalanche documentary. My first thought when I saw the televised images of the World Trade Center collapsing in September 2001 was that the dust cloud boiling through the canyons of New York City's streets looked exactly like the powder cloud of a large avalanche. Powder blast packs a knockout punch not only because of winds that can top two hundred miles per hour, but also because the cloud contains fine-grained snow particles and thus has a density roughly eight times greater than air alone. Powder blast can turn transmission towers into pick-up sticks and rearrange steel railroad bridges. When working my way through Doug's required "hit list," I'd made a pilgrimage up one mountain valley to see a Ford Econoline van that, though parked 150 feet away from where the snow debris came to rest, had been carried 30 feet through the air by powder blast. It was wrapped around the stout trunk of a cottonwood tree in such a way that at no point of impact was the van thicker than four inches, and as Doug had described, the front and rear bumpers were "kissing each other." The steel frame of the undercarriage had been snapped off as cleanly as if it had been cut with a welding torch. The tree was still standing, which as Doug avers, "says something about the strength of cottonwoods."

In May 1975 two men and two boys were on a sight-seeing and bear-hunting trip up the Twentymile River, not far from Portage, when they heard an avalanche that sounded like a "jet plane." The slide charged to the river, killing one man. The rescue report reads:

> Wind velocity at the river was enough to bodily pick up three people and hurl them through the air approximately 50 feet, to blast the bark off mature cottonwood trees for 20 feet above the ground, leave the hull of a fiberglass air boat sheared off 6 inches above the keel line, the sides rent, torn and strung out for 100 yards across the flats. Solid core mountain hemlock 12 inches in diameter along the slide path was splintered and sheared off by the wind . . . the air boat looked like someone had touched off six or eight sticks of 90 percent dynamite with the charge placed in the middle of the boat.

At Williwaw the trees became more battered and scraggly as we ambled toward the mountain until we reached a clearing covered by an enormous apron of snow. From the air we'd been able to take photographs that captured the debris pile in one frame, but on the valley floor even a 28-mm wide-angle lens was powerless against the pile's dimensions. In less than a minute, more than a billion pounds of snow had bowled a mile downhill. That's nine zeros—or, the way Doug thinks, the equivalent weight of almost 1,500 DC-10 airplanes. Before he allowed himself to sleep that night, Doug would calculate that a fleet of nearly 75,000 ten-yard dump trucks would be needed to move the same amount of snow. The resultant debris pile was more than two thousand feet wide and five hundred feet long—an area equal to 22.5

football fields. Just to reach the top of the mound, which had an average depth of twenty feet and sections thicker than seventy feet, we had to climb the height of a two-story house, over boulders of snow the size of cars. Magnus Magnuson, an avalanche specialist from Iceland, later determined that the volume of snow debris exceeded the annual catch of capelin—a mainstay of northern waters—harvested by the entire Icelandic fishing fleet. There was enough snow to stretch across the United States from New York to California in a foot-wide, foot-deep stripe.

As the helicopter lifted us toward the top of the mountain, its shadow lost in the deep crease of Williwaw's track, something was changing between Doug and me. I can't pin it to a definitive moment or a specific comment. I can't even be sure how much I thought about it at the time, though I remember sensing the difference that afternoon as I let more and more feet of safety line pay through my hands with Doug at the other end. I could feel the tension and responsibility of his weight, but I couldn't see him. He had backed off the edge of the fracture line and vanished. Even big slabs in most parts of the world aren't much thicker than twelve to twenty feet, and most of the avalanches that catch people are between one and five feet deep. Slabs break perpendicular to the slope, so as Doug descended into Williwaw's 47-degree starting zone, he spun in the air as though he was dropping into a cave five stories deep. When his cramponed boots finally touched bottom, he stood with his back to a wall of snow that loomed thirty-six feet above him and stretched farther than he could see in either direction—a distance longer than six city blocks.

The sheer breadth of this avalanche was so humbling that it forged a tacit alliance between Doug and me. Already, over those first probationary months of winter, he had taught me much of what I knew, and his skepticism had begun to melt. But the excitement of this day transformed Doug to committed mentor just as I had swilled enough information to appreciate how much I needed one. Most of my life, I'd learned by absorbing the written or spoken word. Now it was as though I had been pushed inside the tiger cage at a zoo, clutching a briefcase full of tiger statistics. The briefcase wasn't nearly big enough to hide behind, and my theories about the way the tiger should behave were only going to take me so far. To stay alive, I had to hone my powers of observation and place a greater premium on experience than is necessary in more forgiving environments. Of course, runs of luck long enough to see me through the inevitable bad decisions that would yield experience wouldn't hurt either.

Doug and I began roaming the mountains together more and more often, seeking out avalanches the way others might tour the latest exhibits at a museum. Snow became a common bond, a shared curiosity and delight. If we ended up at dinner after a day outside, we talked about little else. The fine points of a single weak layer could keep us absorbed through dessert. One night Doug told me that I had the potential to become one of the top avalanche specialists in the country— if I would only do two things. "What?" I asked, practically lunging across the table. "Learn to eat spicier food and drink more beer," he said, with a smile that spread like sunlight.

Neither Doug nor I could have forecast that our relationship would quickly meld into a partnership or, much more slowly, metamorphose into romance. The first time I'd laid eyes on him, when he was giving a public talk in a large auditorium, I'd noticed the glint of a wedding ring on his left hand as he waved his arms for emphasis. Falling in love with him never seemed like a possibility. Not only was he married with kids; he was too old. And he was such an avalanche guru that it didn't feel any more appropriate to fall for him than to date my doctor—or Davy Crockett or Abe Lincoln, for that matter. Ironically, we were so passionate about enhancing our understanding of the union of terrain, weather, and snow variables that makes avalanches possible that we were blindsided by the parallel union of circumstances joining us at the heart.

Within a year I was teaching at Doug's avalanche school, with no compunction about borrowing the other instructors' lines or even their jokes. I look back and wonder at Doug's trust in turning me loose when I knew so little, but he understood that the fastest way to learn is to teach others. When he trimmed the allotted time for my snow metamorphism lecture from an hour to thirty minutes, I couldn't fathom what information to cut, so I talked twice as fast—a windup toy on speed. Now I can give the guts of the same talk in three minutes. Together, we wrote a well-received book focusing on the essentials of evaluating avalanche hazard. My identity began to be as firmly rooted in avalanches as Doug's. My unfortunately distinctive voice was on the hotline nearly every day, and I showed up often enough in the media that before long I couldn't make a trip to the grocery store—or even wander through the Seattle airport—without being recognized as "the avalanche lady."

Sunburst. Baldy. El Dorado. Eddie's Secret Mountain. Each accident or distinctive avalanche became a bookmark by which we'd remember the seasons. In the course of rushing off to rescues and clambering over fracture lines, Doug grew into the best friend I'd ever had. We'd laugh, and debate, and race each other on skis back to the truck before saying good-bye and returning home to separate lives, Doug to his wife and three daughters and me to a series of boy-friends. Doug had loved his wife since high school, once dropping everything to hitch from Boston to Seattle to change her mind after receiving a letter in which she broke off their three-year-long engagement. I popped cheerily in and out of their log cabin home—sometimes for dinner, sometimes to babysit the girls, sometimes to stand in the doorway chatting with Pat while Doug stomped around try-ing to find errant belongings.

But Doug and his wife grew in separate directions in their twenty years together. Several years into our friend-ship, Doug told me he was leaving home and asked if he could camp on the land I'd bought in the mountains above town. By then I was living by myself in an apartment over a bingo parlor while getting organized to build a house. Doug needed, he said, to wake up "hearing the birds sing." When falling temperatures drove him indoors, he slept on my couch. Before long we were sharing a bed. Come spring, ex-cavation began for the foundation of my house. I arrived at the construction site one evening after work to find that the lumber I'd ordered had been delivered and stacked in a tow-ering pile. Gamely, I buckled on my crisp new carpenter's belt, walked a few circles around the heap, and dissolved into tears. I had no idea how to even begin framing a wall. Min-utes later Doug pulled into the driveway towing a trailer full of his power tools.

The enterprise of building a house can be so stressful that it has the reputation for rupturing even the most solid partnerships. As my home began to take shape, however, so, too, did the romance between Doug and me. Still, if not for Doug's persistence in both swinging a hammer and scaling the walls of my heart, I'd probably still be living in an unfinished shack, alone.

At twenty-seven, I knew what I knew, why I knew it, and why it had to be so. I prided myself on being rational and left little room for shades of gray in my thinking. I dated men my age who had gone to prestigious schools. Left alone, my mind easily reduced the romantic sparks between Doug and me to ashes. *The guy is on the rebound. Hell, he's almost forty; he's probably having a midlife crisis. I was eleven when he married, thirteen when he had his first child. He'll be seventy when I'm in my fifties. It's been a flattering fling but* we *don't belong together.*

And yet . . . in the grocery store, I found myself buying more bananas because he loved them. The days I didn't see him dragged. I loved the spark in his eye and his quiet competence, the way he'd stop work to watch two eagles glide overhead or answer the croak of ravens with a raucous call of his own. I loved the way he seemed to be able to devise a solution to any problem. At midnight it was Doug who was up on the ladder next to me stuffing insulation into the ceiling. When I'd get overwhelmed by all that remained to be done before I'd be able to curl up with a book in front of the woodstove, in a house with switches that turned on lights and toilets that flushed, Doug would put a strong hand on my shoulder and remind me to keep building one nail at a time. When we were together, my heart was soothed by a peace and sense of collaboration I'd never felt before. When we were apart, my brain would clang into lockdown mode and I'd resolve to tell him it was over. *But maybe,* my ever-rational brain would interject, *I should wait until the spruce siding is on.*

The day came when the house was almost finished, and

I realized I couldn't imagine living in it without Doug. Perched on a cardboard box in the living room, I admonished him to consider the house his as much as mine. A few days later, I went to town for a quick round of errands. Somehow, in that brief absence, Doug managed to construct such an elaborate workbench for himself in the garage that there was no longer room for my car. When I protested, accusing him of clandestinely precutting the boards and biding his time for an opening, he gave me a piteous look and said innocently, "But you told me to think of it as my house."

❈ MY DEVOTED East Coast mother was less than delighted with my obsession with avalanches and may have been equally dubious about my relationship with an older, recently divorced man. About the latter, at least, she was circumspect. But she has yet to overcome the impulse to ask, on viewing photographs of us sticking our fingers into fracture lines on steep mountainsides, "Is that safe?" What she worries about, of course, is the nearly vertical wall of unsupported snow above our heads. The answer we give her begins with Silly Putty.

Like Silly Putty, snow can flow and bend into ribbons and folds, and it can also bounce or spring. But when yanked rather than gently stretched, Silly Putty will break. If snow is stressed too much or too rapidly, it also becomes brittle and ruptures into pieces. If my mother still looks confused, we whip out an inch-wide, ten-inch-long heavy-gauge rubber band, pull it taut, and point it at her. "Picture this rubber band as the snowpack," we urge, eager to be helpful. When stress is added, say, in the form of a storm—add two inches of new snow, then six, then maybe a little wind loading—the rubber band is pulled progressively tighter. The rubber band stretches viscously and elastically until it is simply stretched too far—and then it snaps. Snow's ability to store energy elastically keeps us employed because it enables fractures to be propagated across the slope and thus makes slab avalanches possible. While Mom has yet to appear grateful for snow's unusual mechanical properties, when the rubber band is stretched to near maximum and vibrating with energy, her level of comprehension seems to skyrocket. Holding her ground, she asks again about the wisdom of meddling with a fracture line. "For one thing," we answer, "the slope has been relieved of a huge amount of weight, that is, stress. More important, once the snow has avalanched, we are dealing with a dead rubber band." To underscore the point, we triumphantly display the limp, broken rubber band, its elastic energy spent. When none of these explanations mollifies my mother, as is generally the case, I have found it best to change the subject.

One vibrant spring afternoon four years into my avalanche career and months into our romance, Doug and I inadvertently tested this neat theory with our lives. We'd begun the day with a simple plan. A friend of ours who worked for the railroad would go up in a helicopter stocked with explosives, which he intended to use to incite avalanches while the railroad tracks below were kept empty of trains and people. At a more leisurely pace, Doug and I would follow in a second helicopter, Sherlock Holmes and Dr. Watson come to investigate the fresh, unsullied "crime scenes."

"I have no data yet," said Holmes to Watson at the beginning of the "Scandal in Bohemia" case. "It is a capital mistake to theorise before one has data. Insensibly one begins to twist facts to suit theories, instead of theories to suit facts." Studying fracture lines is about getting the facts. It is about trying to understand how the terrain influenced where the avalanche broke and why the slide ran as long—or as short—a distance as it did. It is about coming to know the peculiarities and penchants of different combinations of slabs and weak layers. It is a naked glimpse of the wound that results when the snowpack loses the balancing act between stress and strength. The patterns we squeeze from the data over time are useful in forecasting the sizes and shapes of avalanches to be expected from certain slopes under certain circumstances, which can be critical for siting or defending power lines, transportation corridors, and buildings. We can dig ten snowpits and theorize about what might happen, or we can dig into a fracture line to learn what did happen. Doing hundreds of fracture profiles has helped us distill the information most important to pass on to others.

It should have been a red flag that, by a quirk, the pilots of the two helicopters could not communicate with each other over their radios. But we considered this inconvenience no reason to change our plan on a day when, given even slight provocation, the slopes were discharging model-gorgeous avalanches at eighty to one hundred miles per hour. Exhilarated by the show, Doug and I waited across the valley during the strafing runs. Only after the concussive *booms* and *whop-whop-whop* of the receding helicopter had echoed into silence and the clouds of powder drifted back to earth did we take to the air.

Intent on fingerprinting the slopes that had just proven their guilt by avalanching, we landed on a rocky knob near the top of the mountain and began unfurling the ropes we'd need to rappel over the seven-foot-thick fracture line. From the helicopter we'd seen that this avalanche wrapped far around the mountain. Once confident that we'd built anchors strong enough to hold our weight, we clipped ourselves into the ropes and began backing over the brow of the mountain toward the fracture line.

Doug had sailed over the rim of the fracture and I was standing at its top edge when we heard a familiar, increasingly loud *whop-whop-whop* and the first helicopter banked from around the corner, approaching us at eye level. We waved, certain that our friend could see us in our brightly colored parkas and assuming he was taking one last look at his handiwork before heading home. The helicopter flew out of sight, and ninety seconds later, our assumption shattered in an awful *boom*.

The size of the explosive charges used for avalanche control depends on the stubbornness of the snowpack. Ski patrollers commonly use two-pound charges in their winter-long campaign to make the slopes safe for the public. That day our friend had needed nothing greater than a ten-pound charge to command submission from the slopes. But now, with the twin objectives of dumping his remaining explosives and forcing loose a triangular tooth of "hang fire" that had not released in the first avalanche, he'd dropped a fifty-pound charge no more than 150 feet from us.

Cus D'Amato, legendary trainer for heavyweight boxing champions Floyd Patterson and Mike Tyson, once said, "Fear is like fire. It can cook for you. It can heat your house. Or it can burn you down." Fear cooked for Doug. It made him shimmy up the rope hand over hand and vault over the cliff-like fracture line. He had just landed next to me and we were turning to run uphill when the snow began to pitch and jiggle. If the existing fracture line collapsed or a new avalanche broke between us and our anchors, the ropes tethering us to the mountain would likely fail. I looked at Doug to gauge his reaction but saw only myself reflected in his sunglasses.

Even as I fled, I was struck by the tenderness in his voice when he urged speed. With a jolt of recognition that made me run even faster, I realized that he was more concerned about my safety than his own.

We hadn't been lying to my mother. Absorbing the shock, the fracture line bucked and pranced as though in the grip of a major earthquake, but in the end the snow stayed fixed to the slope and did not avalanche. Our unprofessionally passionate hug of relief would undoubtedly have embarrassed Sherlock Holmes and Dr. Watson, but it felt just right to us.

CHAPTER 3

Unburying the Past

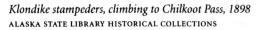

Klondike stampeders, climbing to Chilkoot Pass, 1898
ALASKA STATE LIBRARY HISTORICAL COLLECTIONS

To treat your facts with imagination is one thing,
to imagine your facts is another.
    —John Burroughs, U.S. naturalist

BY THE TIME I MET HIM, DOUG'S KNOWLEDGE OF THE LAY of several mountain ranges had become so intimate that when flying in a helicopter, he could close his eyes and peg the location by the turns the pilot was making. Though he couldn't keep track of mundane possessions like his hat or car keys, he seemed to have a rap sheet for each avalanche path imprinted in his brain. When we drove the Seward Highway, he might take one hand off the wheel (and, more disconcertingly, both eyes off the road), gesture enthusiastically toward the runout zone we were hurtling past, and say, "See the old timbers sticking out of the mountain? That's all that's left of what used to be a snowshed protecting the railroad. Back in 1924 a foreman named Shanklin was trapped inside the tunnel for fifty-seven hours. The rescuers found his hat in the avalanche debris and took him for dead." I'd twist in the passenger seat, still trying to spot the timbers, as Doug accelerated into the next bit of avalanche lore. Sometimes, as we stood atop one of the mountains behind Anchorage, Doug seemed oblivious to the beauty around us. He'd wave his arms from one slope to another, painting a tragic panorama of the fatal avalanche accidents that had transpired within view.

As a newcomer to avalanches, Doug had found the comments stirred up by every accident as striking as the slides themselves. Typically, the implication was that the recent slide was an anomaly—a fluke of nature that had never happened before and was unlikely to happen again. Hearing the same disclaimers about nearly every major avalanche made them harder and harder for him to believe, and he thought that they were giving shape to dangerously solid barriers of misperception. By the mid-1970s Doug concluded that if he was ever to understand avalanches in Alaska well enough to help prevent accidents, he must know their past.

Stretching more than twenty-two hundred miles east to west and over fourteen hundred miles north to south, Alaska encompasses 365 million acres, an area equivalent to one-fifth of the landmass of the contiguous United States. It is more than four times bigger than Germany and six times the size of the British Isles. One glacier alone (the Bering) occupies an area three times larger than the state of Rhode Island. Obviously, Alaska has plenty of the two required ingredients for avalanches—steep terrain and snow. (The record winter snowfall for one site in the Chugach Mountains—Thompson Pass during the winter of 1952–53—is 974.5 inches.) But when Doug combed the card catalog at the University of Alaska's library, he netted only three references to avalanches in Alaska. If he was to unearth evidence of Alaska's avalanche past, he was going to have to hunt for it—in yellowed diaries, exploration accounts, police blotters, hundreds of pounds of dusty railway logs, photographs, and the memories of old-timers.

A reference librarian suggested that Doug begin with

Alaska's newspapers and steered him into the microfilm section. There, Doug was dazed to find more than a hundred old tabloids from Alaska and Canada's neighboring Yukon Territory stored in a series of waist-high file drawers stretching seventy feet. "As soon as I walked into the room," Doug told me, "I realized that this was going to take me a lifetime." The newspapers were not indexed by subject. Doug could only slip a reel into the microfilm machine and begin to read.

Alaska's avalanche history is in its infancy compared to the centuries of chronicled human habitation in the mountainous regions of Europe. Strabo, the Greek geographer born in 64 B.C., wrote of avalanches as "enormous layers capable of intercepting a whole caravan." Even earlier, in 218 B.C., Hannibal's army was likely slammed by snowslides during its epic crossing of the Alps. An avalanche near Disentis, Switzerland, in 1459 destroyed a church that had stood for 655 years. By contrast, the oldest written mention of avalanches in Alaska that Doug has discovered dates to 1790, when the explorer Vancouver sailed into Prince William Sound. The oral histories of various Alaska Native groups include the occasional avalanche. For example, circa 1845 a tower of blue ice avalanching off a glacier into Disenchantment Bay, near present-day Yakutat in southeast Alaska, is said to have produced a wall of water that engulfed a summer sealing camp and killed roughly a hundred Tlingit Indians.

When Europeans arrived in what is now called Alaska in the mid-eighteenth century, it was occupied by around thirty thousand Natives. As Alaska's population ballooned with explorers, hunters, and prospectors pushing into the region, so did the number of ambushes by avalanche. Recorded incidents pick up in the 1880s and early 1890s, mostly around Juneau, in southeast Alaska, where syndicates were mining hard rock gold deposits and prospectors were panning the creeks. But the Klondike Gold Rush is unquestionably a demarcation line.

GOLD! GOLD! GOLD! GOLD!
68 Rich Men on
the steamer *Portland*
STACKS OF YELLOW METAL!

—Extra edition of the *Seattle Post-Intelligencer,* July 17, 1897

In the summer of 1897, two ships reached Seattle and San Francisco loaded with more than a million dollars' worth of gold dust and nuggets—what Northwest Indians called "the yellow metal that makes the white man go crazy." Their arrival sparked what historian Pierre Berton in *The Klondike Fever* describes as "one of the weirdest and most useless mass movements in history." More than a million people—doctors and hoboes, hookers and hustlers, policemen and bankers, saloon keepers and clergymen—laid plans to go north and get rich. Of those, at least a hundred thousand actually left home. Fewer than forty thousand would reach the boomtown of Dawson, and by the time they straggled in, every inch along the Klondike's gold-bearing creeks had been long staked.

Despite the assurances of guidebooks circulating widely by the fall of 1897 that "in a sense all roads lead to the Klondike," there were no easy routes north. Those who could afford to took a steamer from a Pacific Northwest port across the stormy Gulf of Alaska and into the Bering Sea to the village of Saint Michael on Alaska's west coast. They then traveled by wood-fired stern-wheelers more than fifteen hundred miles from the braided mouth of the Yukon River upstream to Dawson. But this circular route was slow and passable only in the short summer months. As described by historians David Neufeld and Frank Norris in *Chilkoot Trail* (the source of the quotes herein), most chose to travel a more direct sea route to southeast Alaska through a tangle of waterways known as the Inside Passage. When the ocean ended in a wall of mountains, stampeders had to climb into Canada through the narrow gaps of either White or Chilkoot Pass and, once in the interior, build boats capable of negotiating five hundred miles of swift water from the Yukon River's headwater lakes downriver to Dawson.

Roughly twenty-two thousand stampeders elected to tackle 3,550-foot Chilkoot Pass, which, though higher and steeper than its rival, offered a shorter route. The trail began at sea level in Dyea, a brawl of a town that mushroomed and died within two years and of which little remains but the Snowslide Cemetery. One of the hopeful, recalling his arrival in December 1897, said, "With people wrangling and fighting over freight, with confusion, great avalanches booming down the mountain sides all about us, and absolutely no one able to give us anything but abuse, my first view of Dyea was accompanied by one long and two thousand short blasts of profanity." Famed naturalist John Muir, disgusted by the same chaos in nearby Skagway, compared the craze to a "nest of ants taken into a strange country and stirred up with a stick."

One guidebook called the Chilkoot route "the meanest 32 miles in history." Stampeders—or argonauts, as they were also known—described the last thousand vertical feet up to the pass on the American side as "almost perpendicular."

With an angle of 35 degrees, it was "like climbing an icy stairway to hell." At the summit, North West Mounted Police enforced a requirement that every person coming into Canada be supplied with a year's worth of food and equipment—at least a ton of goods. (They also lugged machine guns to the summits of both Chilkoot and White Pass to bolster their claim to the region, since the United States and Canada were wrangling over national boundaries.) The average outfit for one person included four hundred pounds of flour, fifty pounds of oatmeal, thirty-five pounds of beans, and twenty-five tins of butter, and this was only the beginning of the list. Stampeders could move their gear by sled or pack animal to the Scales, a tent colony at the foot of the last climb, but from there the steepness of the slope forced them to haul everything on their backs. Unless they could afford to hire human packers at scalper rates of a dollar per pound or managed to avail themselves of a crude limited-capacity tram, shuttling their outfit to the top of the pass took twenty to forty trips over as long as three months. A virtual city of mounded gear accumulated at the pass, and cast-off items— trunks, bedposts, and framed pictures—began to line the trail, sacrifices to fatigue.

Throughout the winter of 1898, an unbroken, grunting human chain extended whiskers to tail up the treeless snow-covered slope. Bent forward by the wind and heavy loads, stampeders of every conceivable age, size, and nationality shuffled in a common rhythm that came to be known as "the Chilkoot Lock-Step." Trenches were cut to the side of the trail to allow spent climbers to rest, though it might be hours before someone who stepped out of line found an opening to slip back in. Stampeders paid the entrepreneurs who maintained the trail a daily fee to use the "Golden Stairs," fifteen hundred rough steps hacked out of the snow, but this was far from the only toll exacted.

Hunched over the microfilm, Doug followed the winter of 1898 in the *Dyea Trail,* the *Daily Alaskan,* and other chronicles. Throughout February and March, one snowstorm after another made it impossible to climb beyond the Scales on most days. At the top of the pass, Canadian Mounties often needed a sentry shoveling throughout the night to keep huts and tents from collapsing on sleeping policemen. Those stampeders who managed to flounder to the top of the pass were unable to find their freight, which was buried under more than twenty feet of new and drifted snow. The argonauts' journals were replete with weather woes. Louie May from Anaconda, Montana—stuck with his party of twelve at Sheep Camp, the last camp in the timber—wrote of his attempt to travel the four miles to the pass on March 22: "Very stormy and I was not in favor of going up at all but the rest think that it is not stormy on summit so away we go and the further we go the worse it gets and when we get to the scales we find one of those storms which you read about, raging over and around the summit." May opined in a separate entry, "There are several women in camp, among them one doctor. I would not want my wife out on such a trip, it is too rough and in many other ways not a trip for a woman."

Doug had grown accustomed to searching for avalanche squibs in side columns and small print, but there was no need to squint when reading of the Chilkoot Trail disaster of April 3, 1898. The Palm Sunday avalanches were trumpeted in tall headlines, not only in the breaking news of the *Dyea Trail* but as word of the tragedy was raced on rival steamers and fanned to the rest of the world. Next to the strike itself, the avalanche deaths became the most widely reported story of the Gold Rush.

Sepulchre of Ice and Snow

AN AWFUL AVALANCHE
BURIES A GREAT HOST OF PEOPLE.

It Came Without A Moment's Warning.

The Saddest Incident Not Only of Great Rush to Gold Fields But in the History of Alaska and the Northwest Territory

Faced with six feet of newly fallen wet snow and warm winds, the local Chilkat Indians who worked as packers grew increasingly fearful, and by Saturday, April 2, they refused to climb to Chilkoot Pass, no matter how high the wage. Some of the Natives stayed at Sheep Camp, while others retreated all the way to Dyea. Two respected longtime (sourdough) guides were also warning anyone who would listen about the likelihood of avalanches, but a break in the weather on the afternoon of April 2 was too much for impatient stampeders, who saw their prospects for wealth diminishing with each delay. Once again argonauts began trudging uphill, a thin black line against an indifferent wall of white.

The lull in the storm was what Doug calls a "sucker hole." By Saturday night those sleeping fitfully in drooping canvas tents at the Scales heard avalanches cascading around them. A slide at 2:00 A.M. buried about twenty stampeders; another slide hours later caught three more. All were rescued, including Pennsylvanian Anna Maxson. Bruised and cold but not terribly hurt, she calmly told her husband of her premonition that she wouldn't live to see the Klondike, exacting a tearful promise that he wouldn't abandon their quest for gold. Terror was creeping over the camp, but when a large group convened in one of the few rickety buildings, the argonauts couldn't decide whether to remain at the Scales indefinitely or to try to descend to Sheep Camp. In any event, no one advocated moving before daybreak.

The first deadly slide let loose midmorning, killing three men, who were later found lying "vault-like" in their sleeping bags. A work crew, days away from completing a new elaborate aerial freight tram, was hit by the second slide—though it would be hours before anyone realized the nineteen men were buried. They had been among the first of more than two hundred people to evacuate, and as the workers had hurried through the Scales from the summit, they'd handed a 200-foot rope to the prospectors massing to flee. Six of the strongest stampeders put themselves at the front of the rope to bulldog through the drifts and deep snow, while others grabbed hold and the two women in the crowd were lashed on. Upon beginning their escape around 11:00 A.M., with long-handled shovels their only known luggage, the group was immediately beset by small slides. The largest slide of the morning struck just before noon, as the prospectors entered a ravine about a thousand vertical feet below the Scales.

"Every time I read a new account," says Doug, "I felt as though I was wallowing blind, hip-deep in snow, barely able to breathe or hear against the force of the gale. Those guys in the lead must have felt like they were trying to punch their way out of a pillowcase." The big avalanche charged through the front of the line, dumping forty-foot piles of snow debris and pitching bodies every which way. The bedlam was such that "it's no wonder," says Doug, "that no two stories quite match."

One man freed from the snow recounted, "The voices of

the entombed reached me from all directions. Many seemed to be praying and some were muttering goodbye to relatives at home. I did not suffer any pain. My agony was mental." Quickly, the snow grew quieter.

No one has been able to determine definitively how many people were rescued or even how many died. Some of the names found on faded wooden headstones in the Snowslide Cemetery at Dyea are not on any of the published lists of victims. One name that appears in both places, however, is Anna Maxson's. Her premonition was only hours old when she succumbed to her second avalanche of the morning. The best guess is that ten to thirty people were extricated alive and sixty-seven killed. At the time it was the greatest avalanche disaster in the United States. More than a century later, it remains the second deadliest avalanche day in North America, runner-up to the avalanche that killed ninety-six people stranded on a railroad line in Wellington, Washington, in 1910. The continent's third worst avalanche accident occurred during the same storm, killing all but two of the sixty-four-man crew working to clear the Canadian Pacific Railway line through Rogers Pass.

On the Chilkoot Trail, more than a thousand prospectors labored two and a half miles uphill from Sheep Camp to begin a confused search. One of the young men dug out was New Yorker Arthur Jappe, who had been given up for dead and left on the floor of the new tram's powerhouse. The *Dyea Trail* reported that when Vernie Woodward spotted Jappe's body, "she cried and begged for him to come back to life to look at her. She unfastened his shirt, and in frenzied grief began to rub him. She worked upon him as only a true woman will—moving his arms this way and that; pressing his chest and breathing into his lungs until three-o-clock in the morning. She was then rewarded by his opening his eyes and speaking her name, Vernie." Romantic songs have been written about the "Lady of the Chilkoot." The only niggling detail is that Jappe had not known Vernie before, and the two never saw each other again.

For the most part, the rescue is a grim story of one body retrieved after another, most frozen as though running, the victims' arms thrown up over their bearded faces. In black-and-white photographs of the scene, the rescuers look like twins of the victims: the latter only happened to be a day ahead on the trail. The bodies were sledded to a tent, where men working for the notorious gangster Soapy Smith stripped them of their jewelry and cash. During the search the trail was closed for four days. At first stampeders worked in rare collaboration, but as time uncoiled, they became increasingly anxious for the blockade to be lifted, and only martial law held them in check. The trail reopened five days after the avalanche and, as though coursing through a floodgate, prospectors again began climbing.

Ask Doug for a tally of April 1898, which stands as Alaska's most infamous avalanche month, and he will answer without pause, as though you have asked his name: "Seventy-two gold seekers were killed. Another forty-nine people are known to have been caught or buried. Tons of equipment and supplies were lost. Thirteen mules and ten dogs died. Another twelve mules, one dog, and one ox survived. The dog's name," he will add, "was Jack and the ox was called Marc Hanna." Two days after the Chilkoot avalanches, Marc Hanna was found under the snow, calmly chewing the bale of hay that had inadvertently saved his life by creating an airspace. Without ceremony, he was put to work hauling the bodies of the less fortunate to Snowslide Cemetery.

Jack was part of a dog team toiling up the Valdez Glacier, supposedly an easier all-American alternative to the Chilkoot Trail, but really a twenty-five-mile nightmare of crevasses, blizzards, and avalanches seldom attempted today. Prospectors gathered for their assault at the end of Valdez Bay in Prince William Sound, near the present-day terminus of the Trans-Alaska oil pipeline. Six miles inland the glacier rose in a series of six benches so steep that block and tackle were required to hoist goods uphill. Canvas tent camps of several hundred people sprang up at the bottom of each of these benches. During the last week of April, it snowed so much that the six-foot-long wooden steering poles taken from the sleds and set on end disappeared completely. At the base of Bench 6, closest to the 4,800-foot summit, the bivouacked prospectors were almost out of wood and oil. Whatever fuel they had, they dared use only for cooking and melting snow for water, so the sole refuge from the cold was to curl up in sodden sleeping bags. On the night of April 30, an avalanche roared over at least twenty tents, entombing twenty-four men and one woman. Others in the camp rushed into the wet whirl of the storm, most clad just in long underwear and some without shoes. They dug madly, not knowing that the rescue effort would save their own lives—a second slide came down while the search was still under way and crushed some of the empty tents. Avalanches claimed more lives elsewhere on the glacier during the same storm, but remarkably, at Bench 6 only two of the buried victims died—both Freemasons from the "Chicago tent." When all the missing had been accounted for, Jack's owner, known on the trail as "Shorty," began digging for his four dogs. The first dog he found was dead. Shorty dug along the trace from its harness until he came to the corpses of his second and third dogs.

Shorty lost hope for Jack and returned to camp. Eight days later, as a group of hungry men probed the debris, hoping to recover some of their lost supplies, they heard a weak yelp from beneath them.

Here it is best to pick up the story in the words of Neil D. Benedict, a flatlander from Florida who was on the trail at the time. For Doug, finding Benedict's unpublished manuscript, "The Valdes and Copper River Trail"—of which there were purportedly only three copies in existence—was like being handed a canvas sack full of gold nuggets.

> [Jack's] interests were neglected for eight days, which he passed in sad reflection on the fickleness of human love, doubtless imagining in his ignorance of the true cause of his imprisonment, that he was being studiously punished by his master for some entirely unknown lapse of duty on his part . . . [Jack was dug out alive] whereas his dogship was of course becomingly thankful; but somehow he was not overly demonstrative in protesting the old underlying affection for his very deliberate master. But after a few days of regular rations, Jack returned to his work on the trail apparently none the worse for wear.

Once lured north, stampeders too disappointed to return home pushed farther into Alaska, rummaging through remote corners. Most never found wealth; in *North Country Challenge,* Ernest Patty (president emeritus of the University of Alaska, who as a newcomer to Alaska's mountains was "cut down to size by an avalanche") recalls that he asked a grizzled prospector where he would go if he could quit his claims and leave the country. Glaring, the man replied, "Young fellow, I'd go someplace where I didn't have to wear shoe-pacs and where they don't call dried apples fruit." The lucky uncovered gold, copper, coal, even oil. But Alaska's newfound mineral wealth could not be exploited without punching transportation corridors through difficult terrain. So began a new chapter in Alaska's avalanche history, one that does not appear to have an ending.

By 1900 the larger steam-powered aerial tram that began carrying freight to the top of the Chilkoot Pass late in the spring of 1898 was being dismantled, made obsolete by a new narrow-gauge railroad running through White Pass. In typical style, the Chilkoot put up a fight. According to a February 1900 issue of Skagway's newspaper, the tram crew "beat the tram to the bottom." The men were launched "over a thousand feet in less time than it takes the ordinary railroad hand to say his prayers." The White Pass and Yukon Railway connecting Skagway to the Canadian Yukon River town of Whitehorse was Alaska's first railroad, albeit with only twenty-two miles of the line on American soil. It remained Skagway's only link to the interior (other than trails) until a road was built in 1978. Bad weather and avalanches from an intimidating posse of avalanche paths known as the "Dirty Dozen" have dogged the railway since the beginning; headlines like STORM KING BATTLES WITH THE RAILROAD PEOPLE FOR SUPREMACY are legion. The triumph of the line's completion belonged to Michael Heney, a brash young Canadian engineer whose battle cry was, "Give me enough snooze [snuff] and dynamite and I'll build you a road to Hell!" He would prove to be a key figure in Cordova's history as well, and his name was lent to the mountain range that launches avalanches into the 5.5 Mile neighborhood where Jerry LeMaster used to live.

✳ TAKING OUR CUE from the Klondike stampeders, Doug and I shouldered our packs in the late summer of 1986 and began to climb the worn ribbon of Chilkoot Trail toward Canada. We, too, had reached Skagway by venturing north up the Inside Passage from Seattle, except that Doug had paddled a kayak while I had rowed backward beside him in an eighteen-foot oceangoing shell equipped with a sliding seat and nine-foot, nine-inch long oars. The 1,400-mile labyrinth of singing humpback whales and headwinds—our first long journey together—had taken seventy-two days and left us lean and fit. A few hours into the hike, we took a break at the Snowslide Cemetery, where the ox Marc Hanna had hauled victims from the 1898 Palm Sunday avalanches. Inside the rickety wooden fence, some of the headboards were so worn by weather and time that we could recognize the names only because Doug already knew them.

Doug had by then prospected almost four thousand avalanche events affecting people in Alaska. With his big beard, he looked like a relic from the previous century, and he sounded equally authentic. "I came to feel like I knew *cheechakos* [newcomers] and sourdoughs alike," he says. "I knew who fought with whom, who mushed to such-and-such a place, and who was particularly down on his luck." Doug is a master of incidental details—if you want to know the story of the two prospectors who hauled the frozen corpse of an avalanche victim across glaciated mountains (and then demanded $500 compensation) or how much cordwood was needed to keep the steamboats running, all you have to do is ask. His stories open like Russian dolls—one is nested within another—and he tells them straight from the heart.

Through the lens of his research, Doug has seen explorers and surveyors, trappers and dog mushers, reindeer herders and loggers, construction workers and linemen, mountaineers and snowmobilers and skiers all get caught, often in exactly the same places. If you ask what year the fore-man of the mountaintop telephone relay station on Mount Susitna stepped out of his house, broke off a cornice, rode thirty-five hundred feet, and survived to quip to the *Alaska Sportsman*, "The next time I empty the garbage, I'll wear a parachute," Doug will tell you 1961. Without glancing at his records, he can rattle off the especially big years in Alaska—1898, 1917, 1932, 1959, 1979, 1980, "and 2000, of course." He is prone to quoting philosopher George Santayana: "Those who cannot remember the past are condemned to repeat it." Yet he contends that he has only "scratched the surface" of Alaska's avalanche history. "On any given day," he says, "I can go out and find mention of a new avalanche with only a few hours' work. And while I'm looking at the past, I can't keep up with the epidemic of accidents happening in the present."

In 1980 an avalanche swept an aging blue sedan off the Seward Highway. Convinced he was about to be "squashed," the driver—an unemployed former oil field worker—dove out a window and managed to stagger on knees, feet, and el-bows across the snow debris, which though still moving had slowed significantly, braked by the much wetter snow at lower elevations. Like a lava flow, the avalanche continued to ooze the last several hundred feet downhill to Turnagain Arm and onto the railroad tracks. The avalanche occurred at 5:45 P.M. in January, which at 61° North latitude means that it had already been dark for almost two hours. Better visibil-ity wouldn't have allowed the engineers of the Alaska Rail-road train enough warning to react anyway, for as soon as the gleaming blue-and-yellow locomotives pulling fifty-two freight cars swung around a bend, they plowed into the still-shifting mounds of avalanche debris. The lead locomotive, weighing 265,000 pounds, rolled off the tracks like a wounded elephant. Three more followed suit. As momentum drove the train more than a hundred linear feet into the rubble, freight cars began to tip and crumple in an awful screech of metal. Dumping their contents, a total of thirteen cars were forced off the tracks before the train came to a stop and silence again took control of the night.

The avalanche occurred only a few miles from Doug's cabin in Bird Creek. When he went to investigate the next morning and found the crew still with the train, Doug asked them if they knew of similar accidents along that stretch of track. "Oh, no," answered the lead engineer, who had worked the same section for fifteen years. "This has never happened before." Skeptical, Doug asked the engineer if he knew of any "old-timers" who might have information. And that's what led Doug to the door of a small white clapboard house that had quite literally had the department stores of modern-day Anchorage grow up around it.

"I don't remember too much about avalanches except for one thing," Bert Wennerstrom told Doug in a quavery voice. He had worked on the railroad in 1920, as a boy of sixteen. Crossing to the bookshelf with measured steps, he pulled down a worn diary and thumbed to an entry marked "Continuation to diary of Wed. 28th [April 1920] about the snowslide at Milepost 75. Told by old Tom of the Section Gang who was one of the survivors."

> It was 12:30 when we started to work and between that time and the time of the big slide which occurred at 3:30 pm, as close as I can judge, 4 other slides had come down from which we had to run for dear life . . . There was about 25 of us strung out at one place when the watcher hollered and pointed up the track. Now there's a deep cut

through the rocks to a depth of about 40 feet where we were and we started hell-bent-for-election up the track towards the way he pointed. And good Gods, we ran smack into the damn slide . . . The force of the snow falling . . . had thrown me and some of the other guys about 30 feet out into the inlet. Now the fellow that froze to death, he couldn't swim and we that got safe could not get at him or the other fellows that were thrown out for about 1 hour. Some swam in while others were rescued. I could hear some of the poor devils that were caught scream and holler for help but we either couldn't see them or find them.

Wennerstrom had penciled a notation at the end of this entry: "While old Tom was telling me this he shook his head and made slight motions with his hands. The engineer on one of the engines had saw Tom get hit. First he was buryed [sic] then he bobbed up only to disappear again. He was rolled for thirty feet that way. It sure was some mess I'll bet out there."

About a year after his conversation with Wennerstrom, Doug received a phone call from a curator at the Anchorage Museum of History and Art. A photograph album that had just been donated by a family in California included some avalanche photos, though she didn't know where or when they had been taken. One look and Doug felt as though old Tom was whispering into his ear, for there were pictures of stranded men on drifting ice floes, even some of the rescuers paddling the bodies back to shore on rafts improvised from ladders.

❄ BY THE TIME Doug and I stood together in the Snowslide Cemetery, we were no longer gainfully employed. Plummeting oil prices had caused the state to lose $220 million in revenue for every dollar drop in the per-barrel price. As it tightened its financial belt in May 1986, Alaska had begun wholesale elimination of social programs. Several legislators sent Doug and me home with instructions to choose between his education program and the forecast center I ran. We sat in our living room asking each other, "So, honey, which program do you want to cut: yours or mine?" We have had bigger debates about what to eat for dinner. To me, the obvious solution to keeping people safe in the mountains lay in equipping them with the skills to make their own decisions. In the end, the legislature spurned our advice and, with the single stroke of a pencil, killed both programs. No matter that the state had the highest per capita number of avalanche fatalities in the nation, higher than the ski mecca states of Colorado and Utah or the moisture-laden Pacific Northwest. Gone was the weather and avalanche hotline for those venturing into the mountains. Doug's internationally renowned avalanche school was also thrown on the slag pile. Alaska now trades with Colorado for the highest total number of avalanche fatalities in the United States, and has since the winter of 1992–93, a staggering statistic given Alaska's relatively sparse population.

As we contemplated unemployment the spring of 1986, the trees were just budding and the mortgage on my house was barely months old. Although we had painted the walls, our dining-room table was still a plywood Ping-Pong table, complete with net. But my name had long since joined Doug's in conversations about avalanches. Every year I couldn't wait for winter to announce itself with freckles of frost across the windshield and a thin tendril of woodstove smoke spiraling into the sky. Sniffing out unstable snow was my favorite pastime, and a week wasn't complete for me unless I'd managed to hang over a few fracture lines. Still, with the few remaining avalanche-related jobs filled, I probably would have moved on to a career outside of avalanches. But Doug said no.

Doug had concluded at an early age that if he didn't stand up for himself, no one else would. When an obstacle is put in his way, he starts climbing. The state legislature might be powerful enough to eliminate the avalanche programs, he reasoned, but it couldn't pencil out the problem. If we no longer had jobs, we would make new ones.

Three inches of paperwork later, we had created a non-profit corporation called the Alaska Mountain Safety Center, with which we shared our home address and a spare bedroom. The state—which had no use for the copyright of our book, the ten thousand avalanche photographs we had taken, the Alaska Avalanche School, or the row of four-drawer metal file cabinets full of Alaska's avalanche past—unceremoniously dumped the lot into our hands. While the paperwork sifted through the bureaucracy and the midnight sun ruled the skies, we shut the door on the clamor of house projects and the insecurities of our economic future, spent five days steaming to Seattle aboard the state ferry, launched our boats, and began paddling north.

A fairy dusting of snow was crowning the peaks when we slipped home to Anchorage in early September. Almost more work than we could handle was waiting at the door like a pack of hungry dogs. One of the first tasks was to publish the season's schedule of avalanche classes. Over the years the hundreds of workshops offered by the Alaska Avalanche School had gained the reputation of being the best and most intensive in the country. Phone calls from mothers had begun cramming our answering machine even in the full bloom of summer—for mothers' worry knows no seasonal limits. Wherever, whenever we held the first avalanche safety workshop of the winter, they would make sure that their sixteen-year-old sons attended.

Typically, boys that age live for snowboarding, the winter equivalent of surfing, and flock to the backcountry, where they are free of lift fees and ski area rules and regulations. Known as "shredders," many speak a unique dialect—a *spore,* for example, is a "stupid person on rental equipment," and *tonar* is "totally gnarly," which is a good thing. Their voices may have dropped into the octave of men, but they are still too young and eager to think themselves mortal, hence the mothers' anxiety. Many of the boarders take their inspiration from movies in which those fearless enough to have become icons leap from cornices and cliffs, cutting daring routes down the flanks of pristine mountains. Those same movies also contain some of the best available footage of inadvertently triggered avalanches, but watching someone else get thrashed only heightens the thrill.

One year three teenagers calling themselves "the Shred Brothers" slouched into one of our workshops. They were there only because it was the lesser of two evils—they had been arrested for skiing out-of-bounds at a local ski resort, and the judge had offered to waive heavy fines and erase their misdemeanor convictions if they attended avalanche school. By the last day, they had become converts, refusing to allow any of their buddies to ski with them "until they got educated." All three have survived into middle age.

Mothers don't fill the answering machine tape by themselves. Wives also tend to call before the season's schedule is printed. At any given one of our three- or four-day avalanche workshops held high in the mountains, at least half of the participants will sheepishly raise their hands when we ask how many have come at the behest of the women in their lives.

Regardless of sex or age, whether ski bum or heart surgeon, everyone comes to avalanche school for more or less the same reason. They want the freedom to pursue whatever passion has enticed them into the mountains, and they want protection. We promise that we can teach them enough to stay alive but also warn plainly that attitude and overconfidence can kill them. The choice is theirs to make.

When Doug started the school, the backcountry was the domain of climbers, skiers, and snowshoers; snowboards had yet to be invented, and snowmachines didn't have enough horsepower to get their riders into much trouble. Almost no one carried shovels, beacons, or even avalanche cords, the low-tech predecessors to beacons—essentially a string with arrows pointing toward the wearer in the hope that the trailing end might float to the surface in an avalanche. (The cord usually became hopelessly entangled in the debris or spooled around the victim.) The prevailing attitude was that avalanches came with the turf, and there was nothing to be done but hope for the best. Doug was convinced that he could change the way people thought about traveling in the mountains. The key to preventing accidents, he believed, lay in teaching people to make decisions based on hard facts and observations rather than on fate, ill-defined assumptions, hopes, and desires.

At the first Alaska Avalanche School workshop I attended—as a student and the newly minted director of the Alaska Avalanche Forecast Center—I drove the seventy miles north from Anchorage to Hatcher Pass in the Talkeetna Mountains, chugging up a thin switchback of a road into an amphitheater of peaks and valleys. The stars shone like jewels on a swatch of black velvet, and faint green curtains of northern lights fluttered overhead. At the berm and parking lot that marked the end of the plowed road, I swung on a pack loaded with warm clothing and four days' worth of food, and skied by headlamp the last mile to a structure barely visible under a roof laden with snow. The building, perched at thirty-five hundred feet, had gotten its start as a gold mine manager's house in the late 1930s, been transformed into a bar in the 1960s, and had only recently been renovated into a visitor center for Alaska State Parks. For the school, it would be a classroom by day (though most of the training would take place on the surrounding slopes) and by night a floor for thirty bodies sandwiched shoulder to shoulder in sleeping bags. The information packet sent in advance had advised bringing earplugs to ward off the inevitable cacophony of snores.

The workshop officially began in the morning when Doug, dressed in a T-shirt and red shorts with holey dark blue long johns underneath, pulled up a chair and began addressing the roomful of similarly half-clad people with the enthusiasm of an evangelist. With their beards, weather-beaten faces, and easy smiles, Doug and the four men he introduced as instructors looked like a band of pirates, but they were setting the sails for a voyage of discovery.

Many workshops later, when I sat in Doug's seat and welcomed participants myself, I could see that many were as nervous as I had been in 1982, though maybe not for the same reason—as a conspicuously new avalanche forecaster, I was more afraid of betraying my lack of knowledge than I was of avalanches. One March a woman sat on the couch by a long bank of windows, her feet drumming the floor. She said, "I don't know what it is about avalanches, but they seem to be gunning for me. Out of a line of traffic on the Seward Highway, it's my car that gets hit and shoved against the guardrail. Five people ski a slope ahead of me, but the slope doesn't cut loose until I'm on it." With an offhand laugh, I told her that she'd come to the right place, but before I could finish my sentence, there was an enormous rumble and the whole building began to shake. A fat slab of snow careened off the metal roof above our heads, thundered onto the porch outside, and tore off the wooden railing. In twenty years of workshops at Hatcher Pass, this holds the record for the most spectacular roof avalanche. It broke only one window—the one directly behind the marked woman. The glass sprayed as it shattered, embedding shards in the woman's hair. The people sitting on either side of her were untouched, without so much as a fragment in their laps. If the woman's eyes had been any wider, she could have walked through them. Gently, we joked about sending her home or having her sign enough liability waiver forms to wallpaper the room, but instead we buried her in the snow.

Picture yourself six hours into the first day. You are outside, with five other participants and an instructor, working on avalanche rescue. You're starting to feel pretty familiar with your avalanche beacon, you now know how to set up a probe line, and who needs practice with a shovel? You're in a huddle with two other people you met this morning, and as you stamp your feet and twirl your arms to urge blood into your extremities, you review the rescue plan you've just devised together. You know that the three other members of your field group and the instructor are around the corner setting up some kind of disaster to which you'll respond. How bad can it be? You've volunteered to be the leader and to interview any witnesses. Alan has been acing the beacon drills, so he'll go after any signals. Kathy, who looks as though she runs marathons as warm-ups for more serious exercise, will scoot around the slope looking for clues. You're ready and you're wondering about lunch; this instructor might be fanatical enough to forget to eat. What could possibly be taking so long? And then you hear a shout. The words are unintelligible, but clearly this is your signal. Your teammates are already sprinting, and if you don't get moving, you're going to be left behind.

Panting hard, you reach the bottom of a slope that has been trampled to resemble avalanche debris. The only person in sight is a woman screaming in a language you don't understand, flailing her arms, and running in frantic circles. "There's a loud signal right here!" shouts Alan. Kathy darts off to check what looks like a ski with a boot attached sticking straight up out of the snow, and you are left to catch up with the crazy woman. She is throwing words at you, but whatever language she is speaking was not offered at your high school. Finally, it dawns on you that she is speaking Russian. "Can either of you speak Russian?" you shout to Kathy and Alan, but they are too preoccupied to register the question. It's up to you to get a hand onto the shoulder of the wild woman and try to understand what she is saying. It does not occur to you to ask whether she speaks English. In fact, she does—impeccably—but in her panic, she has reverted to her native language (and in the spirit of the scenario, she is not going to make it easy for you). You resort to sign language. "How many people?" you ask, holding up one finger, then two, then three. At three, she gets very excited. "Are they wearing beacons?" She either doesn't know what a beacon is, or she doesn't know what you are asking. You take your beacon out from inside your jacket and point at it. She nods yes and no. "Where did you last see them?" She gestures all over the slope. Alan stumbles over your feet, still tracking the signal. The instructor, a drill sergeant/mountain guide kind of guy who has climbed Denali something like forty times, has appeared out of nowhere to bark, "Four minutes down. Hustle! People are dying." Kathy yells, "I've got someone," and you pray that whoever it is speaks English. Alan trips over your feet again and then asks with undisguised contempt, "Shit. Did you forget to turn your beacon to receive? I think I'm tracking you." By the time it is all over, you are pouring sweat and wondering why you are dressed like the Michelin Man. Your team has found the Russian woman's three friends, only two of whom were wearing beacons, in fourteen minutes. "Not bad," says Nick Parker, the instructor, "but I think you could cut your time in half." As your heart ratchets down from a gallop, he wants to know what went right and what could have gone better.

Now it is your turn to set up a scenario for the other group. You think being the hysterical witness might be fun, but Nick volunteers you to be buried. He is boisterously jolly as he helps you dig your own grave. Not until you have donned an extra parka, pulled up your hood, lowered your goggles, crawled in, and let Nick seal the hole do you realize that you are trapped in a dark narrow trench four feet beneath the snow surface, with only the tiny tip of your ski pole indicating your whereabouts. At first it is warmer than being outside and seems a perfect place to take a nap, but then the walls begin to press in and your joints solidify with cold. If this is a "comfortable" burial, as Nick insists—a snow cave really, with plenty of air—then a real burial, with your leg twisted against your back and snow crammed in your windpipe, is too horrible to imagine. People begin walking over your head, and though you can hear them clearly, they don't hear your shouts. It sounds like chaos up there. Why can't they get organized? The stale air smells like dirty socks and exerts the weight of lead. You're wondering why you didn't protest when Nick told you to turn off your beacon, saying he wanted to make the scenario more challenging. At last someone tugs the ski pole, but they are too eager and they pull it from your hand. You know that Nick knows exactly where you are, but still, all you can think is, *Don't leave, don't leave, I'm here.* And then, like bullets fired into water, probes begin whistling through the snow. One hits your leg and isn't pulled back out. Another jab, this time near your hip. You hear cries, and the clink of shovels, and as your hole begins to lighten, the spaces between blocks of snow turn the most gorgeous shade of blue you've ever seen. An arm reaches down through a shaft of sunlight, creating a conduit for sweetly fresh air. You stand on cramped feet, nonchalantly brushing the snow off yourself, feeling heroic and thinking that at least now you know what it is like to be buried for what must have been an hour. Nick looks at his stopwatch and tells the search team, "Nice job. Six minutes."

By noon the next day, your group is high on a ridge with Doug. Skiing uphill with him has been like taking a walk with a dog—he stops to sniff, scratch, and pounce in places that would have escaped your attention. His curiosity seems equaled only by his delight in watching you tune in to the code he sees written everywhere in the mountains. "Nature is continuously kicking out clues that have meanings," says Doug, "but most people live so far removed from nature that they no longer know how to read them." He compares ignoring these clues to crossing a four-lane highway without "listening for the traffic or looking both ways."

Is it safe? The essential problem is one of uncertainty. The key, Doug says, lies in piecing together bull's-eye data that does the most to reduce your uncertainty without getting swamped in a sea of irrelevant information. In the classroom earlier that morning, he had trotted out the well-worn fable about the four blind men in India who come upon an elephant. Never having encountered such a creature before, each man sets about trying to determine what nature of beast is before him. One man, feeling a leg, concludes that an elephant is very much like a thick tree trunk. Another man runs his hands along the tail and decides that an elephant is as flexible as a rope. Standing by an ear as it waves back and forth, the third man is convinced that an elephant resembles a fan. Each man is weighing bull's-eye data but drawing erroneous conclusions. The fourth man, though, appreciating that no one piece of data is as important as the interrelationship of the data, slowly works his way around the elephant until he understands its true shape and dimension.

Doug asks more than he tells. What's the slope angle? Which way did the wind blow last week? Why does the snow sound hollow here? When did this surface hoar form? So what? Why do we care? The questions began while your group was still stepping into skis and adjusting packs. Doug pointed his ski pole toward a steep slope across the valley and asked how stable you thought the snow was to the left of the big rock. *Beats me,* you thought, knowing that the stability of the snow can change within a few steps, not to mention over half a mile and halfway up a mountain. Doug wanted you to quantify your opinion with a number between one and five (the higher the number, the more unstable the snow). You went for a middle-of-the-road hedge with a three, but Doug pressed for the data upon which you were basing your opinion. *Data?* Well, you'd driven to Hatcher Pass in the dark and spent much of the previous day buried. But it didn't look like there had been any new snow for a while, and you didn't see any avalanches on the surrounding slopes. And come to think of it, you hadn't found any obvious weak layers when you'd dug all those hide-and-seek holes the day before. Still, the sand dune–like wind ripples on the snow surface had you worried. Doug also wanted to know how certain you were of your opinion. That was easy. It definitely wasn't anything you wanted to stake your life on.

All morning, while traveling with Doug, your group has been unable to provoke a reaction from the snow. All indications by lunchtime are that conditions are stable. Feeling confident in a wide valley or on a safe ridge is one thing, but it is quite another to plunge onto a slope that you have now measured with an inclinometer and know is 38 degrees— prime stomping ground for avalanches. As you try to peer over the convex edge of the slope, you feel like a very pregnant woman who can't see her toes. All you want is a yes or no answer to the question "Is the slope safe?" Columnist Anna Quindlen has written, "We are a nation raised on True or False tests." As far as you can tell, most avalanche questions are answered with "It depends." You are placing more weight on what isn't in the puzzle you've assembled thus far. You haven't seen any avalanches. The snow hasn't been cracking around you. Temperatures are moderate, without drastic warming. When you push your ski pole down handle-first, it seems to meet equal resistance. Doug has been encouraging you to jump on every small hill or creek bank in sight, but nothing has showed any sign of wanting to slide. The layers in the pits you've dug have looked as homogenous as white bread, and when you isolated columns and applied force, you had to really wallop the columns to get any part to shear. The layers might just as well have been connected with superglue. You don't yet, however, know much about this particular slope except that it is capable of avalanching. Doug looks eager, maybe too eager, to do battle as he waves a snow saw in one hand and a shovel in the other. But this is avalanche school and he is Doug Fesler, so you go.

Leaving a picket fence of skis stuck into the snow on the ridge, your group waddles behind Doug like baby ducks. He has warned you not to venture too far downhill, as if there were a chance of that. You dig more pits, and the biggest surprise is that there are no big surprises; the layers are exactly as you expected to find them. You're beginning to relax when Doug ups the ante. He wants to show you a kind of ultimate shear test. Following instructions, your group widens the uphill wall of one of the pits and saws up the sides and along the back until you have carved a freestanding block that is about twelve feet wide and extends eight feet up the slope. Linking arms, the six of you line up horizontally, like can-can dancers, along the top edge of the block. As you take in the six-foot drop to the pit floor and the dizzying view to the bottom of the slope, you notice that Doug is standing discreetly off to one side. When you invite him to join you, he demurs with a smile. Like a conductor, he raises his saw like a conductor's baton, but then pauses to give one final instruction. Reflexively, your whole line leans forward, anxious to catch every drop of advice. "Whatever you do," says Doug with a sly grin, "don't forget to yell 'Banzai!'"

The plan is that on the count of three, with deep knee bends between each number, your line will leap into the sky and land in unison in the center of the block. You take bets on whether or not the block will shear. You are a human bomb of more than a thousand pounds about to impact a relatively small block that is no longer supported by the slope. Surely, at least a few of the top layers will fail on an underlying layer. The saw goes back up in the air, glimmering in the sun, and together you chant the countdown. Later you will discover that your shout of "Banzai!" (which has no effect on the snow but spurs team spirit) was heard across the valley. For now you're in a cloud of snow, desperately trying to keep your elbows hooked into the arms of your neighbors. It takes a moment for the cloud to settle, and when it does, you realize that you have done nothing except create six formidable backside-shaped craters. As you scrabble back toward your skis, you find yourself wishing the snow were less stable.

By the next morning, you have your wish. It blew hard enough all night to rattle the windows in their panes, the shrieks of the gusts negating even the loudest snores. When you step from the building, a thin crack shoots across the wind-corrugated surface of the snow. You note the drumlike sound of the snow, pause to study where the wind has laid in slabs, look not only for new avalanches but for how deep they broke and how far they ran, and find yourself starting the day with an opinion. And all this before breakfast. You're raring to go until your instructor announces that he's tired of breaking trail, and because he wasn't sure he could find a competent guide, he has hired the six of you. He points to yet another far-off ridge and asks to be taken there—and as you follow his finger, you see that he is sending your group into the heart of avalanche country. There may or may not be a safe route, and it is up to you to find out.

For me, the end of avalanche school was only another beginning. I left the workshop a zealot with a mission. Education, I was convinced, was the solution to the avalanche problem. When I began teaching, I hung on every word spoken by the other instructors during classroom sessions to make sure nothing vital was omitted. If worsening weather made it impossible to see or hear during field sessions, I would keep trying to shout insights to my students long after the other instructors and their groups had retreated inside for intelligible conversation and hot chocolate. If graduates got caught in avalanches, I took it personally and wondered what I had forgotten to say. One night a skier who had taken a class the winter before called us at home. He'd just been dug out from a battering ram of a hard slab avalanche that had bruised him eggplant purple from head to ankles. Only his toes, protected by his ski boots, were undamaged. Buried headfirst, he'd been saved by a friend who spotted the tiniest bit of red ski boot protruding from oven-sized blocks of debris. His voice breaking, he confessed that while he'd been trapped, convinced he was going to die, his thoughts hadn't been with his pregnant wife and their two-year-old child. Instead, he said, "All I could think about was, *Oh, boy, I hope Doug and Jill don't hear about this.*"

Running a business out of our home has meant that Doug and I are together virtually twenty-four hours a day. We started out even sharing the same small room as an office, but one phone-intensive day, Doug punched a hole in the wall with a hammer, strung computer cables through it, and laid claims to the adjacent bedroom. The Alaska Mountain Safety Center is now nearly twenty years old; for most of that stretch, we've worked seven days a week in winter and then disappeared together on long wilderness trips from May to September. Working only a few-second commute from bed has blurred the distinction between work and home, making it perhaps too easy to labor through nights and weekends. On stormy mornings, with the phones ringing and avalanches running rampant, I'm sometimes shivering at my computer before I've had the chance to put clothes on.

Over the span of a few hours on any given day, we might function as receptionist, accountant, secretary, janitor, computer technician, boss, teacher, rescuer, forecaster, and bombardier. Though our consulting projects vary each winter, typically we are responsible for managing the avalanche hazard for companies with a threatened mine, power line, or road, which we do through a combination of measures including forecasting and helicopter bombing. One winter in the Aleutian Islands, a hill with a vertical drop of little more than sixty feet produced an avalanche that put eighteen feet of debris against the back wall of the house built at the bottom. The defense structure we designed to be constructed from old wharf pilings and nets literally put food on the table—we were paid in fresh king crab.

All this togetherness may seem a recipe for disaster, but for us, it has been defining. In tight situations we know what the other will do, and if communication is needed, it is often nothing more than a nod or gesture. Each passing season has seemed to fuse our names like melted glass. Now, even in the newspapers, we are often referred to as DougandJill or JillandDoug. It is almost as though we are thought to be one person. And in a sense, maybe at our best we are.

CHAPTER 4

A Walk in the Park

Wind-sculpted snow © JILL FREDSTON

Within a single afternoon, within hours or minutes, everything you plan and everything you have fought to make yourself can be undone as a slug is undone when salt is poured on him. And right up to the moment when you find yourself dissolving into foam you can still believe you are doing fine.
    —Wallace Stegner, *Crossing to Safety*

Life is a gamble at terrible odds—if it was a bet, you wouldn't take it.
    —Tom Stoppard, English playwright

I NOTICED HIM IN THE CROWDED THEATER RIGHT AWAY. HE was trim and handsome, with an athlete's energy and a flashbulb smile that made him seem more a boy than a man nearing forty. As he sat laughing with his friends, waiting for the performance to begin, he kept tipping his wheelchair into graceful wheelies. "Look," I said, nudging Doug. "Isn't that Nick Coltman?" When we had last seen him, he lay broken and freezing, straddling the wafer-thin line between now and before, before and after, here and gone.

Born in England and raised in Australia and New Hampshire, Coltman was a skilled outdoorsman who had traveled the world for adventure and was accustomed to spending fifty days a winter on skis. But on the day he triggered a tiny avalanche with disproportionate consequences, he wasn't exploring the high mountains of Nepal or etching ski turns across the face of a remote Alaskan glacier. He had only gone for a short walk in what was essentially his backyard.

If the day had played out as so many had before, Coltman would have put his dog back into the car after a two-hour jaunt and driven into town to spend the rest of the afternoon with his wife, Maggie. If he could rewrite the day, he might elect not to go hiking, or choose a different route, or succeed in having his ice axe find purchase on the icy slope. If he could rewrite the day, his life would be the way it always had been, and his future would be as he had planned. But though we must pretend otherwise in order to function, none of us can know the script for any day.

I also spied John Stroud in the theater that night. Although they didn't know each other, Nick Coltman and John Stroud could easily have been friends. Both were tempted north to Alaska in the early 1990s. They lived within a few miles of each other. Both were in their mid-thirties when avalanches interrupted their lives. Both loved to hike, climb, and ski the same mountains. Both loved to come home with stories, which Coltman would tell in the quick, ready style of his trade—he is a newspaper publisher—while Stroud, a computer wizard, might need more coaxing. In describing the avalanches that snared each of them, Coltman might flash his easy grin, while Stroud's more cautious smile would likely be lost in his beard. Both, however, would say they were lucky.

❄ SEVEN MILES EAST of downtown Anchorage is the trailhead for Flattop, the most popular mountain in Alaska. If someone scrambles up only one mountain in the state, it is likely to be the 3,510-foot peak, which lies at the accessible edge of Chugach State Park. If you look north from the summit on a clear day, you can see Alaska's highest mountain, 20,320-foot Mount McKinley (also known as Denali), crowding the sky 150 miles away. A procession of increasingly tall peaks and expansive glaciers beckon from the west. Moments from the parking lot, you leave the last "bonsai" forests of wind-wizened hemlock trees and emerge onto open tundra.

Runners flog themselves into shape by using Flattop as a StairMaster, and families picnic on its flanks. Hikers congregate on Flattop's summit to shiver through the longest night of the year in December and to cheer the midnight sun in June. People get married on Flattop, and they die there.

More mountain rescue missions are launched to Flattop and the surrounding "front range" areas of Chugach State Park than to anywhere else in the vastness of Alaska. The frequency of accidents in the front range is often attributed to the density of the surrounding population, and certainly, without people or our trappings, there is no exposure to danger and, therefore, no hazard. Familiarity and accessibility, however, inflate the problem by making us complacent. The mountains don't behave any differently just because they are close to town. "The most dangerous thing," says Jerry Lewanski, superintendent of Chugach State Park, "is that the area is so close to civilization. There's no transition zone. You don't spend hours in your car driving to get there. There's no time to psychologically prepare."

❄ NICK COLTMAN woke on Saturday, November 11, 2000, with no intention of becoming the front-page story in the *Anchorage Daily News* or the alternative weekly newspaper he cofounded, the *Anchorage Press*. He hadn't even planned to take his ritual dash up Flattop with his chocolate lab, Boozer, until he saw the sun—a precious commodity in Alaska. A run of dreary days had given way to a serenely beautiful late fall morning, with no wind and temperatures a notch above freezing. Two days home from Hawaii, Coltman yearned for the snow he might find at the top of the mountain.

Dressed lightly enough to move quickly, Coltman veered off the trail about a thousand feet below the summit and traversed into a network of long, rocky, fingerlike gullies. Asked later why he had chosen the cold, shadowed route over direct sun, he replied with trademark jauntiness, "There are always fewer people on the dark side." To avoid a hiker descending with two dogs, Coltman swung even farther left than usual, though still on turf he and Boozer knew well.

It had done nothing but pour in Anchorage the previous three days. But toward the end of the storm, the temperature near Flattop's summit had cooled enough to chill the precipitation into freezing rain that greased the rocks with ice before changing to dry snow. Strong winds had stripped exposed ridges bare and blown the snow into shallow drifts not much deeper than six inches. In most places the snow was either nonexistent or so thin that, with every kicked step, Coltman's hiking boots stubbed icy ground.

The crux of Coltman's route was the last several hundred feet below the summit where the gully narrows, steepens, and snakes into a bend. He took a few strides into a slightly deeper snow patch to which he took an instant dislike. "I'd seen similar conditions in this spot before," he says, "and always been able to traverse onto a rock ramp and avoid the danger zone. This time I did something I'd never done before. I looked below me and thought, *Ooh, I wouldn't want to fall down that.* I turned toward the safety of the rock, put my ice axe in, and that's when I saw the crack go."

The slab that dish-plated out from under Coltman was no larger than a Ping-Pong table and barely three inches deep. "There was no sound that I remember. At first it was like riding a pillow. Then I started tumbling, ass over teacup, and I knew I was in for it," Coltman later told one of his own reporters.

———

✳ YIELDING TO the temptation of new snow and sunshine, John Stroud skipped work the afternoon of March 28, 2002, to ski in Chugach State Park. With his two dogs and a friend, Skip Repetto, he drove a half hour from Anchorage into the south fork of Eagle River, a majestic valley where houses grip the lower skirts of Alplike peaks. Parking near the end of the road, Stroud and Repetto fastened mohair skins onto their telemark skis for traction and began climbing toward a ridge that cupped three sizable widemouthed snow-filled bowls. Though both men had taken avalanche training before coming to Alaska and recognized that the terrain could produce avalanches under the right conditions, they didn't know that the name of their chosen playground was Three Bowl Path.

Their first downhill run was perfect. For safety's sake, they skied one at a time, each skier disappearing behind a rooster tail of powder. At the bottom, as Stroud and Repetto admired the fluid calligraphy of their tracks, they couldn't help but decide to make another run. This time they chose a different line that led them into the gut of the southernmost bowl.

As with Coltman's accident, there were a lot of ifs to this afternoon. If Stroud and Repetto hadn't tried to squeeze in a few more turns near the bottom of their second run by crossing a V-shaped creek drainage and traversing onto a north-facing slope, they might never have triggered the avalanche. Nor might it have happened if Stroud, maintaining a safe distance behind Repetto, had heeded his friend's warning to stay put because the snow didn't feel right. If Stroud's dogs hadn't dropped into the gully ahead of him, tempting him to follow, there might not have been enough weight in the gully bottom to provide a trigger—or, alternatively, only Repetto might have been caught.

People can grow pretty unstable if kept in dark, cold places all their lives (which is why so many Alaskans take winter breaks in Hawaii). The same goes for layers of snow that lurk in shadowed areas like north-facing slopes. The instant Repetto crossed to a different slope aspect, his ski poles began sinking into a seemingly bottomless layer beneath the new powder. Repetto was perched atop a house of cards. He shouted a quick warning to Stroud and started looking for the fastest way out.

To appreciate the precariousness of Repetto's perch, you must put your head inside the snowpack. Snow appears to be a solid substance, but it is more like lace, a delicate lattice of frozen water and airspaces filled with water vapor molecules. Newly fallen powder is made up of only 5 to 15 percent water, while the rest is air. Even old settled snow packed to the density of a sidewalk is more than one-half air. The undercover agents of change are water vapor molecules and temperature. If a strong temperature difference exists within or between layers, as is often true on shaded slopes early in the winter when the snow cover is thin or during very cold periods, the molecules migrate from the relatively warm top of one snow grain to the colder underside of the grain above. Rather than creating bonds between the grains, the water vapor accretes to make each grain bigger. The trend is to build increasingly large, blocky, poorly connected grains known as facets or, because of the similarity in texture, sugar snow. The longer the temperature gradient exists and the process is allowed to continue, the larger, more sugary, and more persistent the grains.

Faceted snow, in any stage of development, can be a deadly weak layer. In the grains' most advanced form, when they have progressed beyond angled corners and simple striations to form very large hollow cups about half the size of a pea with glittery jewel-like facets visible to the naked eye, they are called depth hoar. As director of the Canadian Avalanche Association Clair Israelson observes, "Depth hoar is like having your crazy aunt come for a visit. You never know when she is going to snap."

Stroud's dogs, malamute-huskies, were no feather dusters—each of them tipped the scales at more than a hundred pounds. The stress of the two wallowing dogs, combined with that of the two men, was too much for the weak depth hoar in the bottom of the gully. It collapsed with a declarative *whumph*. The collapse began around the men's skis and radiated in all directions. If they'd had the luxury to stand their ground and watch, Stroud and Repetto would have seen the snow surface physically drop an inch as one collapse caused another, then another. As the failure of the weak layer under the slab dominoed uphill, it would have looked as though waves were surfing up the slope with the slow, ominous roll of an ocean swell ahead of a storm. The collapses propagated a full five hundred vertical feet before a slab more than two feet deep and a quarter of a mile wide peeled free. Stroud and Repetto might as well have pulled bricks from the ground floor of a fifty-story skyscraper, only to see the building disintegrate.

Repetto had celebrated his thirty-eighth birthday the night before, but now it seemed he was going to die. Clipped at the heels by the onrushing avalanche, he urged speed from his skis, trying to traverse the eddy line that divided sliding snow from stationary snow. He'd almost made it when he was flipped backward into the froth. Twice he tried to pick himself up, and on the second attempt, he stayed upright long enough to cross the eddy line to safety. He looked down toward the creek bottom where snow from the three sides of the bowl was converging and was sure that he had just killed his ski partner. The channel was churning with huge waves of snow that looked like the wildest of whitewater on an unnavigable river. Somewhere in this turbulence, Stroud and his dogs were drowning.

On the opposite edge of the gully, Stroud thought he was far enough to the side to escape the avalanche when a second wave of snow coming from a different direction caught him around the knees. Blocked from safety by the low but smooth and sloping gully wall and caged by ski bindings that did not release, Stroud felt as if he were wearing concrete shoes. As he clawed at the gully's purchaseless walls, the laminar flow of snow debris rose to his waist, his chest, and then his shoulders, drawing him inexorably under, until snow topped his head and locked him into darkness.

❄ NICK COLTMAN hurtled a punishing six hundred feet down a rocky slope that was only marginally dusted with snow and steep enough to be rated expert terrain at a ski area. When he came to a precarious rest, his head was downhill and bleeding profusely. He couldn't take a breath without feeling searing pain. Nor could he move his legs. His dog Boozer was nowhere to be seen. At first Coltman thought he was buried to his waist, but when he saw his legs twisted like pipe cleaners on the snow surface, he thought, *No, I'm paralyzed. That complicates things.*

Coltman knew that death was certain if he couldn't reach the cell phone in the small pack still strapped to his broken back. His hat and gloves had been knocked off in the fall, and most of his backpack's contents—a few extra warm clothes, water, and snack food—had also free-fallen when the torn bottom, patched and stitched together with dental floss years before, succumbed to the rocks. But the cell phone was zippered into a separate top pouch.

Backcountry travelers have increasingly relied upon cell phones to extricate them from emergency situations that could have been prevented by packing good judgment. I recently returned from the nighttime rescue of two skiers carried onto a steep brushy mountainside by an avalanche. They used a cell phone to call for rescue, but when I spotted them from a helicopter with the aid of a searchlight, they were standing and waving energetically. I could see no sign of debilitating injuries and, with no landing zones anywhere nearby, I knew the process of extrication would be hard, slow, and cold. The easiest, fastest solution, I told them by cell phone, was for them to walk downhill. Once they started moving, they reached the road in twenty minutes.

Accustomed to being self-reliant, Coltman hadn't brought his cell phone for safety. Like Boozer, the phone was simply always with him. Usually, he'd call his wife from Flattop's summit, and they would spin their plans for the day.

"I remember only bits and pieces of the accident," says Coltman. "I remember just the first second or two of the fall. I remember going fast. I remember thinking I had to get to the cell phone. I knew that I wasn't going to stay conscious long. I have no idea how I got the pack off my back."

Coltman's 911 emergency call was recorded right before noon, probably less than three minutes after his fall. It is remarkable for its coherence. He told the dispatcher his name, that he'd been caught in an avalanche on the north side of Flattop, and that he couldn't feel his legs. After giving his cell phone number and precise information about his location, Coltman had no choice but to sever the connection to save battery power.

In *Deep Survival,* a book that explores why some people live through ordeals that kill others, Laurence Gonzales writes: "The maddening thing for someone with a Western scientific turn of mind is that it's not what's in your pack that separates the quick from the dead. It's not even what's in your mind. Corny as it sounds, it's what's in your heart." Coltman could not have survived without the cell phone, but for most people it would have made little difference. They simply would not have had the heart to endure being so broken and so cold for so long.

Willing a helicopter into the air, Coltman lay on the slope, growing ominously colder and murmuring, "Oh, help me, help me, help me," to no one in particular. From his upside-down position, he spotted Boozer several hundred feet downhill and called to him. Boozer, his right rear leg cut to the bone, left his own trail of blood as he limped to Coltman's side.

Maggie Balean worried when she didn't receive her husband's usual call from the top of the mountain. Twenty minutes after his 911 call, she dialed his cell. "I can still hear our conversation distinctly," says Balean. "I said, 'Where are you?' and Nick said, 'I've had an accident and I'm paralyzed.'" He explained that he'd called 911 and couldn't talk long because he needed to save the phone's batteries. "He told me that he loved me and then he hung up, and I started going crazy."

The response to 911 calls for accidents in Chugach State Park is sometimes bungled because while there are almost too many agencies close at hand, few are capable of mounting an effective rescue in rough terrain even a short distance beyond the end of the road. Coltman's call, though, was like the single match that sparks a wildfire. By 12:10 Alaska State Parks, the Anchorage Police Department, and the Anchorage Fire Department all had crews in the parking lot at the trailhead, and a request for assistance had been made to the Alaska Mountain Rescue Group, which initiated a phone and pager call-out to gather a volunteer response team. Helicopters were being summoned from more than one direction.

※ JOHN STROUD was entombed so completely that the only part of his body he could move, albeit almost imperceptibly, was the index finger of his left hand. He couldn't even open his eyes. His first reaction was to yell. He yelled for his dogs, whom he knew would find him if they weren't buried. He yelled for Repetto, whom he had seen get caught but prayed had reached the edge of the avalanche. He yelled until he realized that he'd better save his energy and his breath. With little emotion, he said his good-byes and faded into sleep induced by too little oxygen and a glut of carbon dioxide. He didn't even have time for the full weight of sadness for what he was leaving behind to settle upon him.

An avalanche victim stands a much greater chance of survival if some part of his body or a piece of gear that is still attached to him is protruding from the snow. Playing dead may work for brown bear attacks, but it is the wrong strategy when caught in an avalanche. Victims need to put up a fierce fight as they are carried downslope and use swimming motions to stay near the surface—any stroke is worth trying, even the dog paddle. Avalanches most often kill by suffocation, though broken necks and other forms of fatal trauma have become increasingly common as people jump into ever more ruthless terrain. There is air even in dense avalanche debris, but it is unattainable if the victim's mouth and nose are plugged with snow. Even if the victim can draw a breath, his exhalations will begin to make any available air less accessible by coating the snow surface around his mouth with ice. We always excavate carefully around a victim's face to determine whether there is an ice mask, for that is an indication of how long he lived under the snow; thick ice masks are hauntingly disturbing to find.

Poisoned by their own carbon dioxide emissions, most victims begin to lose consciousness within four minutes, which is a good thing, as they will use air at a slower rate. Brain damage may set in after eight minutes. Within twenty-five minutes, half of all completely buried victims die; within thirty-five minutes, almost three-quarters are dead. To have any chance of living longer, victims must have some sort of air pocket, perhaps one created by throwing an arm in front of their face as they were buried. A victim buried faceup rather than facedown is twice as likely to survive because a larger airspace is created in front of his mouth as his head melts the snow. People rarely survive burials deeper than six feet, not only because they take longer to find and dig out (and are thus more vulnerable to both asphyxiation and hypothermia), but also because they have greater compressive weight upon them.

No helicopters were on the way to help Stroud, and no rescuers had been summoned. Stroud's dogs were dead and would not be found for months. His only chance of survival lay on the slender shoulders of Skip Repetto, who at that moment was preparing to flee.

Repetto was worried that the two other bowls could release and sluice into the same gully bottom, burying him before he had the chance to find Stroud. He could see houses below him. They were tantalizingly close, and Repetto calculated that it would take less than five minutes to ski to them and seek help. Even a few people could speed up the search considerably, and he was urgently aware that time was the enemy. Repetto, though, didn't want to "ski off the mountain alone." In the darkest corners of his mind, he knew that if he chased the mirage of help, his friend would die. He turned his back to the houses.

Stroud's only lifeline to the surface was the invisible electromagnetic signal being transmitted by three tiny triple-A batteries inside the avalanche beacon strapped to his chest. The use of avalanche beacons, which were invented in 1968, has burgeoned since the 1980s, paralleling the boom in backcountry winter recreation and a concomitant surge in accidents. About the size of a television remote control, a beacon—or avalanche rescue transceiver, as it is also known—offers the best chance of finding a buried person alive. Even so, the official statistics are discouraging: two out of three people wearing beacons are dug out dead. This ratio is misleading, as we are more likely to hear about the fatal accidents than the close calls in which one friend locates another with a beacon, digs him out, and they go home with little fanfare. But the reality is that even those wearing beacons can break their necks, be unable to expand their chests enough to breathe, or run out of air long before they are shoveled free. Beacons, like cell phones, can even contribute to accidents by making us feel safer. If we have the attitude that we'll be okay if something goes wrong, then we are more likely to act in ways that increase the likelihood of something going wrong.

A transmitting beacon emits its signal in flux lines shaped like the wings of a butterfly. The distance at which this signal can be detected by another beacon switched to receive generally varies between fifty and a hundred feet, though beacon manufacturers offer more optimistic numbers. Repetto thought that Stroud hadn't been carried very far, and he expected to have a slow, tough climb back up the debris-choked gully. He surged with hope when he hustled within range of Stroud's signal in ninety seconds. Flashing red lights on the digital display of his beacon directed him in a gentle arc along one of the flux lines, closing the distance. Within three minutes Repetto was standing almost on top of Stroud's grave.

Sweating and panting, Repetto heaved off his pack and screwed together his ski poles, which were designed to double as an avalanche probe. It can be tough to know what you are probing. Springy bushes, a layer of slush, or muddy ground can have the same spongy feel as a body. This time, though, Repetto only had to probe a few times before he was sure that he was striking Stroud's body three or four feet below him.

Repetto grabbed the small aluminum shovel off his pack and began hurling snow. With a shovel, moving a cubic meter of snow—roughly the equivalent volume of a household refrigerator—takes at least ten minutes. Without a shovel, it takes five times as long. Repetto dug for ten minutes, and when he leaned into the bottom of the cone-shaped hole, worried because he couldn't see any sign of Stroud, he heard his friend grunting. It was an *ugghhh-ugghhh-ugghhh* cavemanlike sound that suggested stress, if not pain—but Stroud was alive! Fueled by emotion, Repetto shoveled well beyond the point of fatigue. Pouring sweat, he dug and dug and dug through soft debris that kept sliding back into the hole. He dug even as Stroud ceased grunting and he couldn't hear any breathing noises at all. He dug until he was standing in a long, almost chest-high trench. The digging was taking too long—fifteen, twenty, twenty-five minutes. Repetto was working too hard to have an exact sense of time, but he knew that with every extra minute of burial, Stroud's chances of survival were plummeting exponentially. Repetto's arms and back burned, and his hope was ebbing.

Stroud's head came into view first, like a dark atoll in a frozen sea. Stroud was unconscious, and he didn't appear to be breathing. From his knees, Repetto reached down, pulled Stroud's bluish face up, and scooped his fingers into his friend's mouth to clear it of snow. Like a balky engine, Stroud sputtered back to life with a rough cough. It would take another ten minutes for him to regain consciousness, and twenty more minutes of deliberate digging to free the rest of his body. If Stroud had been buried even a minute or two longer, he likely would have died.

❋ DOUG AND I live one ridge over from Flattop, less than fifteen minutes of fast driving away from the trailhead parking lot. As is not unusual on a beautiful Saturday, both phone lines at home had rung simultaneously—one call from Alaska State Parks, who contact us as soon as they hear the word *avalanche,* and the other from the Alaska Mountain Rescue Group, with whom we have worked for decades. Both conveyed the same message: *MOVE!* We were told that a fast response was especially imperative because the avalanche victim who had summoned help was buried headfirst to his waist. Grabbing our rescue packs, we stampeded out the door in whatever we were wearing, which in Doug's case was dubiously adequate canvas carpenter pants and rubber boating boots. I was better prepared as I'd just returned from climbing the ridge behind our house. Only when we were in the car did we have time to wonder how the victim could have made a cell phone call if his head was buried.

As a kid, I loved playing the game of telephone, where a message was whispered from one person to another. By the time it came full circle, it rarely bore more than a trace of its original meaning. The initial information about the location and particulars of an accident almost always becomes similarly garbled. You might think you are going to look for a missing hiker at a certain trailhead when, in fact, the mission involves the evacuation of a hunter who has fallen off a cliff in an entirely different drainage. In this case Coltman's message that he couldn't feel his legs had somehow mutated into a headfirst burial. But we didn't know this yet. As we sped out of our valley, we passed two friends who looked miffed when we didn't stop to talk. "Avalanche! Flattop!" I yelled out the window, noting their surprise that there could be an avalanche with snow not even ankle deep on the ground.

A flaming-red Lifeguard helicopter from one of the local hospitals was in the parking lot when we arrived. It had already circled the north side of Flattop and spotted Coltman, but now the pilot was waiting for us to assess the remaining avalanche hazard before ferrying in rescuers. "There's only room for one of you," shouted a paramedic over the clamor of the turning rotors. I nodded a *Go!* to Doug, who knows Flattop's every contour and can run down steep slopes faster than I can. With ducked heads, the two of them bolted toward the open helicopter door. I turned to the rescuer next to me and shouted, "Who?" I didn't know Nick, but his name resonated; it took me until later that evening to register that he called us periodically to get avalanche condition updates for his newspaper's ski column.

The avalanche hazard call was refreshingly simple for Doug to make. All the snow in the gully had slid, stripping the upper slope to a bed of icy rocks. Like Stroud, Coltman had left catlike scrape marks in the ice, an eerie record of his fight to slow his fall with his fingertips after his ice axe was wrenched away. From the broad, level landing pad of Flattop's summit, Doug blitzed hundreds of vertical feet down the mountain, still thinking that Coltman might be suffocating.

If Nick Coltman had in fact been buried, he might at least have benefited from snow's insulating qualities. As it was, lying bareheaded and thinly clad on snow in below-freezing temperatures, he was losing heat so quickly that hypothermia might kill him before his other injuries could. He was also in danger of rag-dolling farther down the treacherously slick slope. Coltman's arms were flung Christlike across the slope, and Doug, the first rescuer to reach him, had to pry the cell phone from his white frostbitten fingertips. Doug says, "The only good thing about the horribly uncomfortable position Nick was in, upside down on a 35-degree skating rink, was that the increased flow of blood to his brain probably helped keep him conscious long enough to make the call for help."

Coltman recalls, "I knew Doug by reputation. I relaxed the instant he introduced himself, thinking, *Oh good, I'm going to be okay.*" But until additional help arrived, Doug could only do so much. Keeping up a constant patter, Doug worked to assess Coltman's injuries, slow the rate at which he was cooling with heat packs and extra clothes, and anchor him to the slope. Doug knew that Coltman's rescue would be measured in hours, not minutes. He says, "I told Nick straight out, 'Look, you're going to have to be real tough and fight like hell to stay alive. Otherwise, you're probably not going to make it.' Really, I was preaching to the choir because I don't think I've ever met anyone with more willpower."

The fire department medics wisely descended more slowly than Doug and arrived ten minutes behind him. Once on site, they were responsible for trying to keep Coltman's injuries from killing him. The Lifeguard helicopter was useful for ferrying me and other reinforcements uphill, but without a winch, it couldn't help with the urgent task of evacuating Coltman—the terrain was far too inhospitable for it to land nearby. Coltman was a long mile from the parking lot and in such a fragile state that he probably would not survive the painful and time-consuming process of being lowered to gentler terrain. Coltman's only chance lay in a helicopter equipped with a sling hoist that could pluck him to safety.

The National Guard's 210th Air Rescue Squadron had been requested shortly after Coltman's plight became known. But as we gently moved Coltman into a rescue litter and onto the more level platform we'd hacked from the frozen slope with ice axes, we heard by radio that the Pave Hawk helicopter was "forty-five minutes out." For Coltman, time compressed into a blur and the details were lost. For those of us watching him grow colder, become less lucid, and struggle harder for breath while Boozer jealously guarded his side, time elongated and began to grate.

The welcome noise of the approaching Pave Hawk rebounded off the valley's walls until it sounded as though an army of helicopters was in the sky. Down below, our team of nearly a dozen rescuers hunkered protectively around Coltman like musk ox. We knew that as the helicopter moved into a hover overhead, the freezing rotor wash would reach a hundred miles per hour and could easily rip us from an icy perch so tenuous that we could barely keep from falling even when standing still. Looking toward the open door of the helicopter from my position by Coltman's head, I saw two pararescue jumpers, more commonly known as PJs, being lowered by winch at the same time. There are fewer than 350

PJs in the United States. PJs can't leap tall buildings in a single bound, but trained to rescue anyone from anywhere, they come closer than just about any other mortals. These are the kind of guys who think nothing of scuba diving and parachuting in the same afternoon, and who wear watches the diameter of Coke bottles on their well-shaped forearms.

The pilot had to hold a high hover to keep the chopper's rotors clear of the slope and to avoid becoming disoriented in the whiteout created by the rotor wash. As the length of cable increased, the two PJs began to spin like figure skaters, with sickeningly fast revolutions. Worse—for us at least—they were going to land right on top of our huddle rather than safely to the side. Most of our team, with heads down, hoods up, and ears deafened by the helicopter's roar, didn't hear Doug's bellowed warning. One rescuer took a glancing blow to his helmeted head from a spinning PJ's outstretched legs, and another was knocked a short distance downhill. It could easily have been catastrophic, with all of us and Coltman cartwheeling down the mountain. "Sorry to drop in on you like this," quipped one of the PJs as he stood woozily to his feet.

The relentless rotor wash transformed the slope around us as though turning the clock back on the seasons, blowing away all the snow except for the denser lobe of avalanche debris. Coltman's litter was clipped to the slender steel cable within moments. Dangling and twisting as it was winched across a cleft of blue sky, the stretcher diminished into a black dot and disappeared into the dark maw of the helicopter. Three hours after taking his last step, Coltman was on his way to the hospital.

By the time Coltman reached the emergency room, he was dangerously hypothermic, with a core temperature of 87 degrees and an irregular heartbeat. If he'd been two degrees colder, he likely would have been in a coma. "I don't remember the emergency room at all," he says. "I'm told that they wrapped me in a warming blanket and that when they began pumping warm fluids through every orifice, I became quite vocal." In fact, Coltman began joking with his army of attendants and speaking in perfectly comprehensible Chinese to Maggie, who hails from Taiwan. (Yes, he was fluent before the accident.) His body, though, was in pieces. He had a host of broken ribs, both of his lungs were collapsed and filled with blood, and his spine was cracked in two places, leaving him without feeling below his waist and rendering him paraplegic.

❊ WHEN JOHN STROUD regained consciousness, even as he shivered uncontrollably, he thought obsessively of his dogs. They had always gleefully joined him, eager to do whatever he was doing. They were family and, since his recent split with his wife, his sole roommates. It tortured Stroud to think that he had brought them into this dangerous place and killed them. He and Repetto, though, were in no shape to do more than hurl unanswered cries into the darkening sky and limp off the mountain.

That night, a shaky Stroud telephoned friends and family. One of the people he called was Don Zimmerman, the only other Alaska employee of the software company for which Stroud works. Stroud described the avalanche in detail, warning his colleague that similarly treacherous conditions must be lurking nearby. Three days later, on a beautiful Easter Sunday, Zimmerman and his stepfather, along with two more dogs, died in an avalanche in the neighboring valley, near a peak named Mount Magnificent.

Stroud's avalanche was orders of magnitude bigger than Coltman's, and yet he walked away essentially uninjured. If

Coltman had traversed only one or two feet higher on the slope, he never would have triggered the avalanche. The slab that caught him was no thicker than a pizza box.

Just as little slabs on big slopes should command our respect, so must relatively large avalanches on little slopes. In Alaska two soldiers were killed on a creek bank with a vertical relief of only thirty-six feet. Two children near Denali National Park—where visitors from all over the world come to marvel at the immensity of the Alaska Range—were completely buried when playing on a hill no higher than fifteen feet. Similar stories abound from avalanche-prone places around the globe and even from milder, less mountainous climes caught unaware by a rare snowstorm.

Every accident, of any kind, is preceded by a chain of events or errors, but each is set into motion at one irreversible moment. Until that moment the accident might have been prevented. Darkness, an icy road, high speed, an inexperienced driver, and a heavy foot on the brake can conspire to create dangerous circumstances, but a car crash might not be inevitable until the car starts to skid. In Stroud's accident the irreversible moment announced itself with a *whumph* as soon as Stroud joined his dogs and his friend in the gully. For Coltman, the irreversible moment came when he stepped onto the slab. Asked in an interview by his own paper, the *Anchorage Press,* whether he'd made a mistake, Coltman answered, "Not really. If I had thought to myself, *Shit, I really shouldn't be going up this,* and I'd gone ahead anyway, then yes, I would tell you I made a mistake. But that's not what happened. The instant I realized I was in a bad spot, I tried to get out of it. Unfortunately, that was an instant too late."

Avalanches are perceived as the equivalent of a drunk driver, barreling through an intersection and nailing innocent pedestrians. Whether small like Coltman's or abnormally big, like the billion-pounder in Williwaw Path, they are frequently met with incredulity and portrayed as freak events. The harsher truth is that as long as people live, work, or play in the mountains, avalanche accidents are certain. Roughly 95 percent of the avalanches that catch those at play are triggered by the victims. These victims are typically experienced, as were Coltman and Stroud, rather than rank beginners. If we can predict that these accidents will happen every year, can we more specifically predict where, or when, or to whom?

When the police apprehend a driver suspected to be drunk, they request that he exit the car and walk a straight line; if he can't, he is considered inebriated. Of course, the test is not foolproof, because some drunks can walk straight. The thin line that tethers us all to life is invisible, far from straight, and famously fickle. It is a line we are constantly walking yet are only allowed to stray across once.

At the rehabilitation center where Nick Coltman was sent to adjust to his new body, he met patients who had become paralyzed as the result of doing nothing more dramatic than wrestling with their kids in the front yard or tripping on the sidewalk. How can we tiptoe through the mountains—or even through everyday life—and manage to stay on the safe side of a line we can't see? How can we stay aware of the thin line without letting fear of it paralyze us?

There is a moment toward the end of some avalanche searches when I feel as though I am suspended above the thin line with a clear, unbroken view of both sides. It's almost as if I can still see the ribbon stretching across the finish line, though I know a runner has just come crashing through. The result of the race hasn't yet been announced or at least spread very far, so it is a very private moment. I had such a flash on the flanks of Mount Magnificent after we'd dug out Don Zimmerman and his stepfather, along with the stiff bodies of their two dogs.

It was a warm day, with sparkling mountains etched into a sapphire sky. With the search over, all the hustle had drained out of our actions, and we'd gained some mental space by zipping the two men into misleadingly cheerful yellow body bags so that their faces were no longer visible. I sat on my pack amid a cocoon of rescue friends and watched a body bag sway beneath the helicopter that was ferrying the victims to a destination of grief. I still knew little about the dead snowshoers except for their names, and I hadn't yet met their families. For a short soothing interlude, while I waited to be one of the last off the mountain, I was free to soak in the sun and let myself be washed by laughter and conversation that was life affirming rather than irreverent. Soon enough this innocent interval would be replaced by the flashing lights of emergency vehicles in a crowded parking lot, microphones thrust into my face, and the wrenching questions of families who wanted to know how the day had gone so wrong. One of the wives would ask me if her husband's eyes had been open and whether I thought he had lived long enough to know he was going to die. Doug and I would go home and try to pick up our lives wherever we had dropped them; by nighttime, I knew, the faces of the victims would intrude upon our dreams. But for now I was at rare liberty just to be.

❄ DON ZIMMERMAN's memorial service was held the week after Easter, on yet another ruthlessly sunny day. From the church parking lot, Stroud could see both the valley where he had almost died and the mountain where Zimmerman had not been so lucky. "I couldn't help but wonder why I survived and they didn't," he says, in a voice husky with embarrassment. "It was a strange feeling. I didn't feel shame. I didn't feel bad. I wasn't happy. Mostly, I was just struck by the oddness of being alive."

A week or two later, at the request of his employer, Stroud went to Zimmerman's house to collect some work-related material. The house was high on the mountainside, commanding the entrance to the valley of Zimmerman's death. "I stood at the picture window with Don's widow's mother," Stroud says, "and she told me over and over again how much she hated snow. And all I could think about was how beautiful the snow looked at that moment."

Stroud normally speaks softly, but his voice pinches to a whisper when he mentions Don's family. "My stomach still hurts when I think how many times I told them that I'd meet up with them to talk about the avalanche. My intentions were good. But I had survived, and Don and Bill had not. Really, I was scared to go back there. I don't know if I had the energy to be supportive."

Stroud feels as though he lost the two months immediately following his burial. "I just missed life for a while. I couldn't remember having specific conversations; I kept losing track of time. It was almost as though I was in a drug state. I'm not even sure why. It wasn't the first time I'd almost died; I bet most people don't think about how close to death they are every time they drive down the highway."

Ironically, the experience of being trapped ultimately offered Stroud greater freedom. "I look more critically at what I'm doing," he says, "and do more of what I think I need." Even more than a year later, the avalanche still runs through his mind several times a day. "A lot of different things went wrong, like the dogs dropped into the gully ahead of me and

I didn't keep my distance from Skip. But enough little factors went right . . . ," he says, turning aside to wipe tears from the pond of his felt-brown eyes.

❄ "IT SUCKS being in a wheelchair," Nick Coltman said without a trace of self-pity when we spoke in his office nearly three years after his accident. "Everything is much more of an ordeal now, even just getting onto the couch or into a car. I get more attention than I deserve. I'm often ushered to the head of a line, even though I can wait as well as the next person—after all, I've got a chair."

From his window, he can see Flattop Mountain, and the accident still fast-forwards through his head all the time. "I'm not bitter or angry; it's just something that happened. I could have slipped in the bathtub or gotten into a car accident. What I miss the most is the crucible of friendship that comes from going out into the mountains with a bunch of friends. My friends miss it too; I think a lot of them are more depressed about what happened than I am." As he spoke, I was reminded of a passage in Wallace Stegner's novel *Crossing to Safety*: "Long-continued disability makes some people saintly, some self-pitying, some bitter. It has only clarified Sally and made her more herself." Nick Coltman radiates that same buoying clarity.

While we talked, an aging Boozer lay dreaming on the floor, and Maggie, also a mainstay at the *Anchorage Press*, could be seen through a plate-glass window. "I've always been fast and lucky," said Coltman, "and able to get myself out of jams. But that day my luck ran out." He paused, creating a silence that was not awkward or even sad. "Actually, I'm lucky as hell. I should be dead. I don't think I could have come back if I'd been any closer to that line."

CHAPTER 5

Rules of Engagement

Turnagain Pass area, Kenai Mountains © DOUG FESLER

The forces we engage are relentless. Gravity is on duty all the time.
    —Laurence Gonzales, *Deep Survival*

Picture yourself at an edge of the world, a slice of sky dangling beneath your skis. The horizon begins and ends with you. Every nerve ending is prickling, and your heart is spurting rich, warm blood. Your legs are springs; your mind is floating free; the mountain is your soul. Deliberately, you try to slow each quickening breath, drawing the air inside of you, washing it through your lungs, and only reluctantly spilling it back into the atmosphere. If you were a rocket ship, this would be the moment of takeoff, and you would rise on a pillar of flame. But you are a skier seeking to meld grace and speed. It takes only one push to give yourself to the snow, and you begin to dance down the mountain in a spray of crystalline white light.

Todd Frankiewicz had subsisted on stale air for months. Aching, he'd sat inside the prison of hospital walls and listened to doctors tell him what he couldn't do with a body that had known no limits for thirty-five years. Todd had been driving home late at night in July when, without warning lights or siren, an Anchorage police officer ran a stop sign and T-boned him on the driver's side, crushing all the ribs on his left side and shoving his spleen and stomach against his heart. He'd even been dead for a few seconds, but in shuffling steps he had come back; he was still coming back. Now it was December and the mountains were whitening into winter. Some people are born with silver spoons in their mouths; Todd was born in Vermont with skis on his feet. To reclaim his freedom, he needed to ski. Tuesday, December 6, 1988, with several feet of new snow and clear skies, was the day he chose for his release.

Though enough muscle had wasted off his body to provoke friends to carefully joke that he looked like Mahatma Gandhi, Todd was feeling better than he had in months. And to have both powder and sunshine in southern Alaska in December was like finding the crown jewels in his morning oatmeal. With two friends—Jerry Steuer and Regan Brudie—who also lived to ski, Todd headed for Turnagain Pass, a little over an hour's drive south of Anchorage. Tested and tuned by several years of pioneering skiing, the men's friendship had been cemented by runs down mountains steeper and more remote than most others dared tackle.

Turnagain Pass is a mile-wide valley bisected by the Seward Highway and framed by peaks that rise from evergreen pedestals. Nothing but a few parking lots and an outhouse at the 1,000-foot elevation of the road flag the pass as one of the most popular winter playgrounds in the state. For skiers, climbers, snowshoers, and snowmachiners, the valley bottom tends to be a starting point rather than a destination. No trail markers point the way; in fact, there are no official trails. Most choose their own routes and begin climbing.

In a normal year Todd was eager for first runs, but this year he made eager look sedate. At home he and his girlfriend, Jenny Zimmerman (no relation to Don), discussed the obvious avalanche hazard. Over the weekend, avalanches choking the Seward Highway had earned front-page headlines in the *Anchorage Daily News,* and a warning from Terry Onslow, the highway's lead avalanche forecaster, had been printed in oversize italics: *"The snow is very unstable. I don't think you'd want to go venturing out on a cross-country ski trip in the mountains right now. The snow is nasty."* Wanting more details, Todd—a friend of ours—called us Monday night. Though we talked snow at dinner parties, he'd never before phoned us at home to ask for a personal update. Doug spent ten minutes describing the nature of the instability in Turnagain Pass while Todd asked careful, intelligent questions. Doug told him that the bottom line was to "avoid steep north-facing slopes like the plague unless they had already avalanched."

That day Todd had spirited leave from work to apply for a marriage license at city hall. After nine years together, he and Jenny planned to marry on the winter solstice, only a few darkening weeks away. At the annual Christmas party held by the manager of the outdoor store where they and many of their friends worked, everyone would be drinking eggnog and obsessing about skiing when Jenny and Todd would step forward, grin conspiratorially, and spring their announcement. And then they'd do the deed on the spot, for Alaska has an uncharacteristically liberal law that allows any resident to perform a legal wedding, a necessity in a state where vast expanses of tundra, mountains, or ocean can separate the romantically inclined from the nearest minister or justice of the peace.

Turnagain Pass offers an expansive menu of slopes, with a variety of angles and aspects. By the time Todd warmed up his truck on Tuesday, he knew that north-facing slopes steeper than 35 degrees were particularly dangerous. Accelerating down the highway, he drove beneath hulking embankments of avalanche debris just cleared from the road. Todd loved steep, deep powder, but he was no fool and he was not unafraid of death. He would stick to the gentler south-facing slopes.

Todd, Regan, and Jerry knew all of the other regulars readying themselves in the parking lot and could tell who had gone ahead by the cars left behind. Many worked frenetic summers as carpenters or fishermen so that they could ski without distraction all winter. The ranks in Todd's group would swell and shrink throughout the day as skeins of friends converged and traded easy banter. Given the avalanche threat, nearly everyone was heading up the broad familiar spine of Tincan Mountain. Todd and his partners availed themselves of the trail—really a trench—that those ahead had broken through the new snow. Where the shoulder of the ridge steepened and left the trees behind for good, they had to step over the fracture line of an avalanche triggered by the lead skiers. Leaning against their poles on breaks, they spotted a rash of fresh avalanches on nearby peaks, most on north-facing slopes. Recent avalanches on similar slopes are the most emphatic clue about potential snow instability. Todd understood this, as did his partners.

Their first run was a blur of laughter and play. Wherever they could, they took turns cutting ski traverses across the tops of avalanche-prone slopes, flexing their knees with enough exaggeration to convey their body weight into the layers, trying to trigger slides. Each time, the snow withstood their method of questioning. Night descends in midafternoon in December, and the close of the day approached quickly. The men were accustomed to doing multiple runs—

"yo-yoing," in skiers' vernacular. The idea of climbing higher on the mountain to seek unscribed powder for a last run was born of exuberance and was so natural and spontaneous that no one explicitly said yes or no, except for a friend who had migrated into the group. On the strength of a "bad feeling," he chose to turn around and descended a mellower route on his own.

The summit of Tincan belonged to the three men on this day, their domain the panorama of ridges, valleys, and slopes beginning to glow pink. Having done the long climb, it made sense to look at the North Bowl, a challenging run that swept from the summit ridge to the trees in one broad, plunging, slightly curving swath. They'd thought it off-limits at the start of the day, but the snow had been proving itself trustworthy. Clearly, the worst of the avalanching had coincided with the storm. As they approached, they could see that the North Bowl had not avalanched and it had not been skied. The wind had tweaked the snow more than on southerly slopes, but the skiing still looked good. Their view was proscribed by the sharp edge of the ridge, but they knew that the run would be steep and fast, the perfect plum to cap Todd's comeback day.

Todd led the way into the bowl that now bears his name, careful to stay high on the slope, not much more than twenty feet below the lip of the ridge. It was a short-enough distance that he must have been confident that any avalanche he provoked would break below him. It took only seconds to execute his traverse and—nothing happened. He shouted an invitation to his partners, and Regan dropped in next.

Just as Regan skied within arm's length of Todd, the snow began to break around them. The fracture line propelled itself uphill with so much energy that it shot over the lip of the slope and would have pulled Jerry off the flat ridge if he hadn't jumped back with the quick agility of a cat. Regan remembers hearing Todd say, "Oh fuck," and watching transfixed as the fracture line unzipped the mountain. And then he was bumping along at great speed on his back, his legs caught up on a large block of slab that was slow to disintegrate. Regan was spit out of the avalanche just above a nasty "meat grinder" of a rock chute. He managed to stand, unhurt. Todd was not as lucky. The avalanche threw him a half mile down the mountain.

Imagine yourself perched on an edge of the world that is shattering beneath your feet. The snow is parting as though cut with a knife, threatening to swallow you whole. Fear will crowd you soon, but for a moment—a fraction of a moment—you surge with exhilaration. The rush folds into panic. You are choking on the mountain; if only you could get one clean swallow of air. Instead of floating like a feather, you are being pummeled like a log in relentless surf. If you see the sky, it is in desperate snatches—mostly you are unable to tell up from down. You are nothing more than skin and breaking bones, a rag doll in the teeth of the mountain.

❄ ALMOST ANYONE who stays in the rescue business long enough comes to regard the telephone as an instrument of torture. It interrupts birthdays and quiet weekends at will, making no concessions to convenience or inclination. On December 6, 1988, the phone rang as we were making dinner after a day spent investigating several of the new avalanches. A skier had been caught in an avalanche in Turnagain Pass. Could we be ready if the Alaska State Trooper helicopter picked us up in the usual spot at the top of our driveway in fifteen minutes? "Jesus, I hope it's not Todd," said Doug as he threw a lid on the stir-fry and flicked off the stove. I didn't bother to answer. Like a child who needs no convincing to believe in Santa, I was utterly sure that Todd wouldn't bite such obviously rotten bait.

I heard the helicopter begin to circle as I sat in the kitchen shoving my feet back into damp boots and was in position next to Doug and our stack of gear by the time it touched down. With a nod from the pilot, we bent low to avoid the guillotine of the revolving rotors, ran forward, and scrambled aboard.

I knew that when I plugged into the intercom, pilot Bob Larson would say with a smile that lifted the reddish-blond ends of his handlebar mustache, "We have to stop meeting like this." For years he'd been the one constant in emergencies, and we'd come to rely on a lot more than his customary greeting. With a slow drawl and Scandinavian roots, he never said much, but he needed fewer words than the average person. In a ground blizzard at midnight, Bob once flew Doug eight miles farther up a glacier than military pilots with night goggles had ventured—and did so as coolly as if he were strolling down a sunny sidewalk. Another night, as he was leaving me alone in rough country twenty-five miles from the nearest road, he motioned me toward his door, leaned out, and shouted into my ear, "Don't worry, Jill, I'll always come back for you." It was a promise he would eventually break.

We skimmed south, catching up on summer news and winter plans. A temperature inversion had lidded the valleys with fog so thick it felt as though we were flying in our own galaxy. Roads and glowing clusters of human habitation vanished, leaving behind only peaks and stars. As we approached Turnagain Pass, we needed help returning to earth. Bob radioed to the Trooper on the ground and requested that several vehicles, with their headlights shining, be arranged in a semicircle in the parking lot that was pinch-hitting as a landing zone. Like a spacecraft, we set down gently amid the diffuse beams of light.

Doug and I ran from the noise of the helicopter to confer with the Trooper in charge. A half step behind Doug, I reached the huddle in time to hear the name Todd Frankiewicz and stopped as though I'd run headlong into a concrete wall. Each forthcoming detail stung like hail. Todd was dead. His avalanche beacon couldn't protect him from being twisted and buried under two feet of snow, but it had allowed Jerry and Regan to locate and free him quickly. They had pumped their own breath into his chest for an hour. There would be no rescue. All that was left for us to do was to bring Todd home.

On the chance that he could land nearby and retrieve Todd's body without ado, Bob wanted the weight of just one passenger. He and Doug flew to the north side of the mountain, scanning with a swiveling spotlight until they saw a jagged crater in the debris. Todd's body lay propped against one wall. The debris was too steep for Bob to set the helicopter down without danger of sliding backward or grounding the rotors. Instead, he toed into the slope with one skid, holding as low a hover as he dared, while Doug crept onto the skid and jumped into the void. Then Bob swung away to wait for Doug to dig a makeshift landing zone.

Alone in the sudden, brutally quiet darkness, Doug switched on his headlamp, and there was Todd lying spread-eagled on his back. "I took off my gloves to close his eyes, which were rolled toward the sky, brush the frost off his eyelashes, and zip his jacket," says Doug, "and then I started chewing him out. It was just me and Todd, and I let him know that I couldn't goddamn believe he had screwed up. What the hell was he thinking? I mean our conversation wasn't even twenty-four hours old. It was like he had been given the key to the castle but had chosen to batter down the back door instead. If anyone had seen me, they would have thought I was nuts. I just stood there for several minutes yelling at a corpse, calling him a stupid son of a bitch in every way I knew how." Even sixteen years later, when Doug remembers this night, his spirit seems to crack. It is not a memory he can suppress at will.

When Doug had shoveled a reasonably level pad, several of us flew in with Bob. With gloomy faces lit in flickers by bobbing headlamps, we worked to push and pry Todd's frozen body into the helicopter's backseat. It could have been a scene from a horror movie.

JENNY ZIMMERMAN's phone was ringing insistently that night as well. The first call was from a friend who had heard the avalanche reported on the evening news and wondered what she knew. Jenny can't remember who called next. The third caller told her that the radio had just announced that the victim was a thirty-five-year-old male skier. Only then did she feel the first tentacles of dread. A friend who had lost her boyfriend to an avalanche two years earlier came to share Jenny's vigil. A neighbor and coworker named Randy, who was nearly the same age as Todd, had also gone skiing at Turnagain Pass that day. "I was standing outside the house," says

Jenny, "when a Trooper car began creeping up the street. It reached the end of the cul-de-sac and began to turn. At that point it could have been either Todd or Randy. And then the Trooper saw me standing there. When he asked, 'Are you Jenny Zimmerman?' I almost fainted."

Later that night, after a futile visit to the hospital, Jenny says, "I cursed Todd's name until I couldn't move off the floor. We'd talked about avalanches all week. When he dropped me off at work that morning, my last words were 'Be careful.' I've never been so disabled with anger."

Todd's parents had long yearned for Todd and Jenny to marry. The day after Todd's death, before flying to Vermont for his memorial service, Jenny made her own pilgrimage to city hall to get a copy of the marriage license as a gift to them. It was a request that melted even the clerk's composure. Supported by a knot of friends, Jenny was pinching all that was ever to be of her marriage between her fingers when another clerk overheard Todd's name. Eager to be helpful, she came forth with the paperwork she had just completed. And so it was that Jenny left the windowless basement office with her marriage license in one hand and her fiancé's death certificate in the other.

Jenny and Todd had already bought special candles for their wedding. The candles burned to stubs at Todd's memorial service.

※ IT'S HARD TO KNOW what to do after manhandling a dead friend. We went home, threw dinner in the garbage, duly answered a frenzy of media calls, and fought passionately about something that neither of us can remember. In bed my thoughts were fragments, my mind the gyre of angst that writer Anne Lamott has described as a bad neighborhood that shouldn't be visited alone at night. One minute I was still walking loops around the Turnagain Pass parking lot, my arm around one of Todd's almost catatonic friends; the next I was remembering the stubborn weight of death as we'd wrestled with Todd's corpse. Next to me I knew that Doug was replaying every word of his telephone conversation with Todd, wondering what else he could have said. He kept rolling over, as if somehow he could find a position of comfort.

We said a lot to each other, even in the long stretches of silence that punctuated our conversation, and sometime during that interminable night, we decided to get married. In the parking lot at Turnagain Pass, I'd struggled to ensure that the Troopers treated Jenny with the respect that would automatically be given a spouse. As "girlfriend," she had no legal standing. In the bitter hours of the morning, it became pressingly crucial to Doug and me to purge our bond of any ambiguity.

※ BEING RATIONAL isn't all it's cracked up to be, especially on no sleep and a stomach that feels full of crushed glass. We probably would have been better off leaving the hows and whys of this accident alone. But habits, by definition, are hard to break, and we were dedicated to the concept that understanding avalanche accidents was the key to preventing them. Just as Todd needed to ski, we needed to investigate the avalanche in daylight. The next morning we didn't have the time to drive to Turnagain Pass and climb as Todd and company had done because a brewing storm would soon destroy the evidence. When the Alaska State Troopers couldn't come up with a helicopter, we prostituted ourselves. We picked up the phone and propositioned the local television station: if they chartered a helicopter to take us and a reporter of their choosing to Turnagain Pass, we'd give them an exclusive perspective of the avalanche.

The lead paragraph of the article on the front page of the *Anchorage Times* that day read, "Three experienced back-country skiers were about to ski a steep mountain face . . . when a mammoth avalanche broke loose and caught two of them, sweeping one to his death." Give me the skeletal facts about any skier, climber, snowmobiler, snowshoer, or any other brand of winter recreationist killed in an avalanche anywhere in the world, and I can write a pretty close facsimile of the article that inevitably appears in print. The stories will trumpet the experience of the victims and extol their skill at their sport. Most articles will imply that the avalanche roared out of nowhere, a total surprise. If the hyperbole is inspired, the avalanche may be referred to as "an act of God," though this phrase is usually reserved for avalanches that destroy houses built at the bottom of avalanche paths. The word "mammoth" will generally be used instead of less terrifying, if more accurate, adjectives like "medium-sized" or "large." Small avalanches are apt to be called "freaks." Avalanches almost always "sweep" or "crush" their victims. Typically, the headlines incite fear, as in DEATH THREAT LOOMS HERE. They can be unintentionally humorous, as in U.S. FOREST SERVICE SAYS SNOW CAUSED AVALANCHE or IT WAS A SNOWSLIDE, NOT AN AVALANCHE. The latter was the result of creative spin by a ski area spokesperson trying to downplay the worry and embarrassment of an unexpected avalanche.

As much as Doug and I shudder with frustration when avalanches are portrayed as unpredictable bogeymen, families and friends are even more likely to scream in pain and anger if the decisions of the loved ones they have just lost are held up to public scrutiny. "I never give them hell," President Truman told *Look* magazine. "I just tell the truth and they think it is hell." If I'd been asked to write the newspaper article about Todd's avalanche, I would likely have led with: "Todd Frankiewicz might not have died yesterday if he hadn't been badly hurt in a car accident last July."

By noon we were standing in dishwater light on the edge where Todd had stood less than twenty-four hours before. Only he was dead, and we were feeling murderous. We didn't have to be crack investigators to get the mountain to talk; it was practically squealing. Our view was the same as Todd's had been, except that the avalanche had cleaned four feet of snow from the bowl and left mostly bare, frozen ground. Swiveling our heads to the south, as Todd surely had, we could see that the wind of the previous week had scalloped the snow surface, eroding snow particles and catapulting them northward. Where the shoulder of the mountain began to slope toward us, the texture of the snow became smoother, gently rippled, and creamy looking. Seen from a distance, the ripples paralleled each other in curves the shape of a drawn archer's bow. An imaginary arrow laid across the apex of the bulges signaled the wind direction, and long, fluted tails extending from the leeward sides of rocks confirmed as adamantly as pointed fingers that the windblown snow had been deposited on the north face. If the texture of the wind slab was classic, so was its hollow sound—the mountain's equivalent of beating a drumroll before battle. All it took to know that the slab was sitting on nothing but rotten snow was a ski pole. When pushed through the more resistant slab, the pole plunged effortlessly through depth hoar all the way to the ground, as though a trapdoor had sprung open. This kind of layering asks for attention almost anywhere, but loaded in a smooth, treeless bowl that holds a 40-degree angle for the first seven hundred vertical feet and has curving side walls barring easy escape, it demands reverence. Todd might as well have stepped in front of a bus. The clues to instability couldn't have been more obvious if lit in neon.

Adding venom to our mood was the reporter. With a dramatic fracture line at his hip and warning signs all around, he commanded a rare stage from which he could say something meaningful and maybe even save lives. But he was one week out of Los Angeles and interested only in his coiffure. He seemed to think that all hope of landing a job as anchorman and escaping the exile of Anchorage rested on having perfectly aligned locks, an ambition being trifled with by the crosswinds raking the ridge. When freezing fingers finally forced him to pocket his comb, we did the interview with snot dripping from our noses and tears swamping our eyes. The piece was flimsy when it ran, little more than scenery and shots of the reporter climbing in and out of the helicopter. His hair, though, looked terrific.

Facts are like building materials. You can pile them up, but they won't construct a house on their own. The terrain, snow, and recent weather explained the avalanche, but they didn't tell the story of the accident. The unsolved variable in the equation was why Todd had allowed himself to get caught.

"What we call 'accidents' do not just happen," writes Laurence Gonzales in *Deep Survival*. "There is not some vector of pain that causes them. People have to assemble the systems that make them happen." Though Todd cannot tell his own story, refrains of it are heard almost every time avalanches and people cross paths. Standing on the edge of one tragedy after another, we have come to know this story so well that it has worn deep creases into our souls. It is a story that has less to do with nature than with human nature. It has many beginnings and fewer endings. Perhaps as good a place as any to begin is with a gorilla.

❄ DO YOU THINK you'd notice if a gorilla walked across the room? This is neither a rhetorical question nor as stupid as it sounds—the subject warranted study by a team of Harvard psychologists. They made a video of basketball players dressed in black-and-white uniforms tossing a ball back and forth and spliced in one of two unadvertised twists. For a full five seconds, either a woman with an umbrella or a person dressed in a gorilla suit sauntered through the action. The psychologists split their study group, asking half of those watching the video to count the number of passes and instructing the other half simply to watch the video. Everyone in the latter group noticed the gorilla or the woman. But 35 percent of those told to count passes did not see the woman. Even more staggering, 56 percent completely missed the gorilla.

Most of us live by the assumption that we only have to open our eyes to perceive what is there. But if, for example, we are expecting to see basketball players throwing a ball *and* we are tasked with an assignment, we necessarily narrow our focus in order to get the job done. The tighter the focus, the more likely we are to miss what may be in the rest of the frame. All of us routinely waltz past gorillas. What we see often has more to do with what we have seen in the past or what we hope or expect to see than it does with what is staring us in the face. I was reminded of this recently when, on a beautiful July evening, one of my closest friends put a homemade pizza in the oven and went for a half-hour bicycle ride. Less than a half mile from home, he flipped over his handlebars and was picked off the pavement a quadriplegic. Now I see wheelchairs everywhere. Presumably, they have always been in my orbit but were not part of my way of knowing the world.

Of course, what we think we see determines how we act. The story of how Todd missed the gorilla on the mountain begins with an age-old motive—desire. Gonzales writes: "The human organism . . . is like a jockey on a thoroughbred in the gate. He's a small man and it's a big horse, and if it decides to get excited in that small metal gate, the jockey is going to get mangled, possibly killed, so he takes great care to be gentle. The jockey is reason, and the horse is emotion." To stay alive, he comments, "We have to use the reins of reason on the horse of emotion."

Todd was riding a whinnying, rearing giant stallion of emotion his first day back on skis. Still smarting from the sting of vulnerability and enforced months of living life slowly, he wanted to feel like his old bulletproof self. He hungered to feel whole, strong, and vibrant in the mountains he loved. He wanted to carve quick symmetrical turns that sent a rush of wind against his face and a plume of snow over his head. He wanted to feel a burn in his legs, and laugh with his friends, and tell stories on the way home, and do it all again as soon as he could. But he could have done all that— and still gone home and played with the dog and married Jenny—if any one of a number of chips had fallen slightly differently.

Although my mother remains dubious, sharing the mountains with avalanches, even at times of high instability, can often be done at low personal risk by reading the snow scrupulously and picking careful routes. Conversely, more than a few victims have created high hazard for themselves during periods of low or moderate instability by exercising poor judgment.

Year after year at avalanche workshops, we have watched participants take in exactly the same information and use it to draw opposite conclusions. During a classroom scenario called "Mythical Pass," we purposely try to squeeze students between the proverbial rock and a hard place—and sometimes the ensuing debates have grown heated enough to trigger lingering marital disputes. Often people learn more from making bad decisions than good ones because the consequences are less ambiguous. The idea behind the scenario is to pay for such understanding with only theoretical deaths and, thus, possibly save actual lives in the future. So, as the class of thirty divides into groups of five, put yourself in one of them and draw your folding chair close.

You are on the third day of a four-day ski traverse through uninhabited mountains. Your group is now standing at the base of a steep but short pass that is the crux of your route. For the last seven hours, you have been traveling up a narrowing valley surrounded on both sides by looming snow-filled slopes. You are bone tired and hungry from breaking trail through a crust layer all day, the weather has been deteriorating for the last five hours, and there is only an hour of daylight left. You must decide whether to hustle over the pass and down to a lower elevation before the storm entirely closes in or figure out an alternative. Your group has only enough food for one more day (an unnamed pariah accidentally left a food sack in the car), one fuel bottle (for cooking and melting drinking water from snow), and the usual gear for a trip of this kind. You don't have a helicopter at your disposal, and in any event the cloud ceiling has dropped too low for flying. Your group estimates that it will take no more than thirty minutes to climb to the top of the pass and another five hours across easy terrain the next day to reach the cars. It looks like the storm could last several days.

What are you going to do and why? As your group gathers data, the instructors are willing to be your eyes and ears, provided you prioritize your questions in the order most likely to reduce your uncertainty. If you don't ask, we won't tell. We'd like the first question to be about slope angle or clues to instability, but inevitably the opening query will be about who forgot the food. "Funny you should ask"—Doug will laugh merrily in response to a question about what the pass looks like, before whipping out a slope profile that a "geologist friend of yours" supposedly drew last summer. If you inquire about the layers in the snow, Doug will ask you if your group dug a snowpit and, if so, where. As the data emerges, it becomes clear to you and most of the others in your group that the slope is plenty steep enough to avalanche, the snow is primed to release either naturally or under the weight of your group, and the weather is only making conditions worse.

Is your group going over the pass? Getting up and over quickly seems to be everyone's initial impulse. But as the data begins to roll in and assumptions take hold, groups typically divide into majority and minority opinions. If you concern yourself only with terrain, snowpack, and weather considerations, you know with increasing conviction that if your group tries ascending the slope from the bottom, you will trigger an avalanche—at best, you will be left with the hopeless task of trying to pull off a rescue in the dark, in the middle of a blizzard. But even if you decide not to go and your group is somehow not too exhausted to backtrack the way it came and feels capable of navigating in whiteout, you are concerned about avalanches nailing you from the side slopes. Camping at the base of the pass where you might get similarly avalanched also doesn't seem viable. Brainstorming, your group looks for alternatives where there appear to be none—and begins to find them. You could retreat to a wider part of the valley and set up camp. However, the powder blast from a fast-moving slide may well flatten your tents and leave you stumbling around in the middle of the night trying to find your gear. Another idea is to back off only a short distance from the base of the pass. You have three or four hours of hard labor ahead of you, but once installed in your snow cave, your group has a reasonable chance of finding both safety and sleep. Even if an avalanche rolls over the top of your cave, you should be able to extricate yourself as long as the cave has a strong, thick, arched roof and you go to bed with your shovel and your probe. You figure you can live awhile with no food and, if need be, can use your body heat to melt snow. If you're lucky, when your group pops out of the cave in the morning, the crux slope will have avalanched naturally and you can be on your way.

If you are really concerned about running out of food and water or worrying your family by being overdue, you are likely to make less of the potential avalanche hazard. You might argue that because conditions are worsening rapidly, it is better to go now. The slope is short, your group will travel quickly and one at a time, and all of you are well equipped and proficient with rescue gear. You can assume the storm will last for days, exacerbating the consequences of running out of food or water. You can assume that the lack of obvious comfortable alternatives means that there are no alternatives. You can rationalize that it is better to be exposed to danger for a half hour than for the duration of the storm.

As instructors eavesdropping on the debate, we often watch one assertive person sway his or her whole group into either staying or going. If they attempt to climb the slope, they will die. We know this because Mythical Pass is based on a 1978 accident that occurred in the Taylor Creek drainage of Turnagain Pass, less than two miles from Todd's Run. A group of ski mountaineers stopped to discuss the obvious avalanche hazard. At the urging of their leader, they continued toward a saddle in the ridge that they estimated to be less than thirty minutes away. We know they made it to the base of the steepest section of slope within ten minutes of their goal because Doug found a water bottle there. Four of the five skiers died. The weather was so bad and the debris so deep that despite repeated rescue attempts, the first body was not recovered for nine months and the last body took twenty-one months to find.

In real life and in hypothetical scenarios, we've repeatedly observed that those with a high tolerance for risk generally filter information about potential hazard and draw optimistic conclusions that lead them to push the fine line even finer. Those who are conservative by nature tend to use the same facts to bolster a conservative approach. For many, Todd's conversation with Doug—combined with recent new snow and wind loading, hollow-sounding slabs, and fresh avalanches—would have been enough warning to stay home or to play on gentle terrain. But, as children know instinctively, the more intensely we want something, the more reasons we will likely find that make it okay. "The trick to forgetting the big picture," writes Chuck Palahniuk in his novel *Lullaby*, "is to look at everything close-up."

Todd chose to place the greatest weight on the fact that his group had been "kicking the tires" on small test slopes all day, ignoring the fact that these tests were mainly done on southerly slopes and the bowl in question faced the opposite way. Like a car buyer already visualizing himself driving the make and color of his choice off the lot, he was becoming less and less inclined to look closely at the paint job or stick his head under the hood. The message jamming his group's circuitry was that incredible—and ephemeral—snow was there for the taking. Most people don't decide to get married after only one date; it took Todd nine years. So why did he put his life on the line on the basis of such a slender thread? Ironically, what may have kicked open the door to trouble was Todd's "experience."

Like most avalanche victims, Todd excelled at his sport. He could ski the way others only yearn to ski and was comfortable in precipitous chutes that would freeze most into statues. With his ability came an unasked-for authority. He was looked to as a leader and often relied upon to make decisions about potential danger. With snow a play pal since childhood, Todd had better-than-average avalanche savvy, but as is also common among victims, his avalanche skills were not commensurate with his travel skills. Presuming that the two go hand in hand is like assuming an Olympic swimmer is a trained lifeguard. Most of Todd's avalanche knowledge was based on what he'd seen, and previous brushes with avalanches had punctuated his learning curve. Todd was a veteran of so many days on snow and enough close calls that if he underestimated the hazard, he was also vulnerable to overestimating his ability to deal with it. In this context, as Gonzales writes, "the word 'experienced' often refers to someone who's gotten away with doing the wrong thing more frequently than others." We think of experience as a classroom, yet it can also be a prison.

In the days preceding the accident, the media sounded enough alarm bells to send a chill through the less experienced. As a powder skier, Todd knew that if he stayed home every time there was a murmur of avalanches, he'd miss the skiing of dreams—and he'd hear about it ad infinitum from those who had ventured forth. With each telling, the snow would grow more and more legendary.

Todd's pick of a sunny day, familiar turf, and similarly talented and gung-ho friends all upped the ante by making him feel not only comfortable, but exquisitely good. The party atmosphere of friends drifting in and out of the group didn't help. As avalanche forecaster Bruce Tremper writes in *Staying Alive in Avalanche Terrain*, "Safety in numbers . . . served us well through the eons when lions, tigers, and bears were our greatest fear . . . Our herding instinct has the opposite effect in avalanche terrain." Bigger groups tend to make us feel safer and act bolder, a dangerous combination. Describing a common phenomenon with unusual honesty, Tremper has told me, "When I'm by myself, I'm very cautious. Add a trusted partner, and I'm willing to go places I probably wouldn't before. Add a group of six people and a couple of attractive females, and I'll do just about anything."

As counterintuitive as it seems, Todd might have been less complacent and, therefore, safer on such an unforgiving day if he'd been skiing unfamiliar territory. In reviewing more than seven hundred avalanche accidents in the United States between 1972 and 2003, Ian McCammon, a robotics engineer turned avalanche researcher, found that 71 percent occurred on slopes known by the victims. As vertical and imposing as the bowl was, Todd had skied it on beautiful late spring days when the snow was stable and his body unbroken. In the inner workings of his brain, he had tagged it a happy, rewarding place. Antonio Damasio, author of *Descartes' Error,* might note that, for better or worse, the slope had subconsciously become a "beacon of incentive."

What would it have taken for Todd to see the gorilla? Maybe if he'd been with partners he didn't trust so much, he'd have steered the group differently. Maybe if Jenny had been there, she could have changed the fateful group dynamic. But even the somewhat unusual act of courage of the friend who opted to be the "nerd" and turn around did not break the momentum. When Todd entered the bowl, he probably still thought he had options. His initial plan was to feel out the snow, and he surely took some comfort in the avalanche beacon he was wearing. That the fracture could rocket upslope of him and tug snow off the flat ridge probably did not occur to Todd because it was outside his experience. The vision of his plan was narrowed by his expectations. "Making a judgment means we create a 'mental model' of an expected universe," explains sociologist Charles Perrow in his landmark book *Normal Accidents.* "You are actually creating a world that is congruent with your interpretation, even though it may be the wrong world."

Atul Gawande discusses fallibility in *Complications,* exploring the effects of experience, wishful thinking, hubris, habits, peer pressure, complacency, inattention, and intuition on the way people dispatch uncertainty and make decisions. His subjects aren't eager outdoorsmen, though; they are doctors. Like it or not, doctors are out in the same weather as the rest of us. Research has shown that the likelihood that a patient will have his gallbladder removed varies as much as 270 percent depending on who is making the decision. A patient with a back problem in Santa Barbara is five times more likely to be operated on than someone with the same back pain in the Bronx. However frightening, embarrassing, or frustrating these inconsistencies, they are normal; they are what make us human.

It is tempting to stay mad at Todd or to label him stupid or reckless, if for no other reason than to pretend we're not just as vulnerable. "Every day," says Dale Atkins of the Colorado Avalanche Information Center, "each of us is an accident trying not to happen." Automobile safety studies tell us that if we drive a mere two miles, we generally make four hundred observations and forty decisions. In the same stretch, we typically make one mistake—maybe we pull up too close to the car in front, or don't stop fully at a stop sign, or forget to flick on the turn signal. Usually, the mistakes are inconsequential, and even if we notice them, they don't seem to merit a moment's reflection. If someone pulls out of a side road and cuts us off one morning, we tend to shake our fists and pretend that we would never be so "stupid," but most likely the day will come when we are in a hurry, or preoccupied, or exhausted, and we will do the same thing. For me, though, such understanding has only come with time. Years passed before I dropped the shield of anger I felt toward Todd, opening myself to a much more penetrating sadness.

Todd's accident can be best explained by asking what he most wanted to do that day—and then answering the question from the mountain's perspective. The reply is a resounding "So what?" Nature's rules of engagement are fundamentally and irrevocably different from our own. Nature doesn't care a whit about what we want to do, or what we think we can do, or how we feel, or what we've done before. As poet Maya Angelou has said, "Nature has no mercy at all. Nature says, 'I'm going to snow. If you have on a bikini and no snowshoes, that's tough. I am going to snow anyway.'" Our assumptions, timetables, needs, skills, and experience make no difference to a hair-trigger snowpack. It doesn't matter that we are jazzed, or that our ego is at stake because we've tried to climb this mountain five times, or that we're afraid to look like wimps in front of our peers, or that we are bolstered by "Kodak courage." In avalanche country, knowing the rules of engagement isn't enough. To stay alive, you have to play by them. Staying alive requires no less than thinking like an avalanche.

Most of the latest crop of skiers and snowboarders who jump into Todd's Run know nothing about Todd's life or death. They are skilled and zealous, and, like Todd, some will leave their names on the slopes where they take their last turns.

For me, Todd's death marked a pronounced loss of innocence. He wasn't the first friend to die in an avalanche—and I didn't yet anticipate that he would be far from the last. A few years earlier Doug and I had been asked by Dan Daugherty, a blond god of a fireman who was also chairman of the Alaska Mountain Rescue Group, if he could tag along with us to investigate avalanches. An avid mountaineer, he was interested in beefing up his avalanche skills. As slides caught forty-two people in less than a month, we had plenty of opportunity that spring of 1985 to climb avalanche after avalanche together. In the same valley where John Stroud and Skip Repetto would fight for their lives, I took one photograph that still gives me pause. It shows Dan kneeling before a fracture line that had just killed a skier and a dog; with his gloved hand, he is sifting the obvious weak layer through his fingers. The next spring Dan was dead. He was killed, along with three others (including the boyfriend of the woman who came to comfort Jenny the night Todd died) on Mount Foraker in the Alaska Range, by a slab avalanche thinner than four inches. From what Doug could surmise from the tracks he saw while searching from the air, one rope team of two triggered the slab, which was perched atop blue ice. Though tiny, the avalanche was enough to knock them off their feet and send them careening into the other rope team. All four men were washed over a 6,000-foot-high wall of ice and rock.

Because of distance and the impossibility of recovering Dan's body, I had been a step or two removed from his death. By the time Todd died, I had investigated enough avalanche accidents to know that, more often than not, those involved were cognizant of key clues before the avalanche occurred. A typical survivor might say: *Well, we knew the slope was steep enough to slide. The snow in our snowpits made us pretty uncomfortable, but . . .* There were always so many buts. *But we thought it was safe because there were already tracks on the slope. But I'd been on this mountain a hundred times before and never had a problem. But we planned the trip for six months and weren't about to throw it all away because of one storm. But I didn't want to be late for work.* Because most of the victims had been strangers, I'd been able to float along on an evangelical tide, believing that more education could solve the problem, my own version of Nancy Reagan's "Just say no" campaign. With Todd's death, I understood from the deepest level of my gut the presumptuousness of fighting human nature.

Doug's innovation in avalanche education had been to expose attitude as an insidious serial killer. His teaching was predicated on the idea that potential victims could be taught to make less subjective judgments. But as the deaths mounted, he had begun to despair of the possibility of keeping people and avalanches apart, and to wonder whether he was wasting his life in a hopeless cause. Todd's death widened Doug's doubt, like an open door that feeds oxygen to a burning building.

CHAPTER 6

The Game of Jeopardy

*A snowboarder catching air above a 50-degree slope,
Coast Mountains, southeast Alaska*   © BILL GLUDE

What does the snail say, riding on the turtle's back? Wheee!
　—Philip Simmons, *Learning to Fall: The Blessings of*
　　an Imperfect Life

IN WORLD WAR II, A RENOWNED SOVIET STATISTICS PROFESsor repeatedly refused to retreat to a shelter during German air raids on Moscow. Economist Peter Bernstein recounts in his book *Against the Gods* that when the professor was asked to explain himself, he reasoned, "There are seven million people in Moscow. Why should I expect [the Germans] to hit me?" One night the professor surprised his friends by descending into the shelter. "Look," he said. "There are seven million people in Moscow and one elephant. Last night they got the elephant."

Statistically, avalanche victims are elephants. Hundreds of thousands of avalanches fall each year, but worldwide—from Europe to Asia, Russia to New Zealand, North to South America—the annual avalanche fatalities rarely tally to more than two hundred. During the eighteen-year stretch between 1985 and 2003, there were 1,756 avalanche deaths in France, Austria, Switzerland, and Italy, the most avalanche-prone European countries. Given the number of people playing, working, or living in avalanche terrain worldwide, it is remarkable that fatalities don't occur every week of the winter; snow can be surprisingly forgiving and is stable much of the time. If, in an average year, we knew of every incident in which people were caught, buried, or injured in avalanches, the numbers probably still wouldn't climb out of the statistical nowhere of the low thousands. Far more people die annually from salmonella or choking on meat than will typically die in a decade from avalanches.

Traditionally, the lion's share of European victims have been residents of exposed villages, while in North America, recreationists have led the charge since the 1950s. This is changing; a pie chart of deaths in both regions over the last decade shows that backcountry recreationists account for almost three-quarters of the dead. The annual number of fatalities is on the rise, particularly in North America, where potential also exists for a dramatic increase in the number of residents caught as homes are built in increasingly marginal terrain. While ski areas host the greatest numbers of skiers, they see less than 1 percent of the fatalities, testimony to the effectiveness of rigorous risk management.

Once in a while, avalanches so huge and destructive occur that even calling them exceptional does them an injustice. The ice avalanche that broke off the summit of 22,205-foot Huascarán in Peru in 1962 was one such beast. It took less than fifteen minutes to drop thirteen thousand vertical feet and charge ten miles, engorging millions of cubic tons of ice, snow, rock, water, and mud. The avalanche annihilated whole villages and claimed 4,000 people and 10,000 animals. Exceptional, however, should not be construed to mean unprecedented or unrepeatable. A similar earthquake-triggered avalanche off Huascarán eight years later killed roughly 20,000 people. More recently, in 2002, a massive break off a glacier in Russia's Caucasus Mountains killed at least 141 people, some of whom lived twenty miles from the avalanche's point of origin. Russian president Vladimir Putin spoke of a "catastrophe of unprecedented proportions," yet similar avalanches had devastated the same valley in 1902 and 1969, and it is a sure bet that they will happen again. Natural hazards offer some of the most infallible odds around. In *Against the Gods,* Bernstein comments, "Mother Nature, with all her vagaries, is a lot more dependable than a group of human beings."

The word *risk* stems from the Italian *rischiare,* meaning "to dare" and implying both opportunity and choice. If we take on risks but treat them as games of chance, we shape our fate. "Risk and time," observes Bernstein, "are opposite sides of the same coin, for if there were no tomorrow there would be no risk." Every day we are blitzed with sensational disaster stories that keep us unbalanced with uncertainty about the bad things that *might* happen to us—though proportionately few ever will. We sit in our living rooms, our doors barred against homicidal strangers, watching news of the latest killer virus across the globe while snacking on enough chips to invite heart disease onto the couch with us. The longer the time lag between taking a risk and feeling its consequences, the more likely we are to ignore the risk. In their eventual impact, longer-term risks like smoking or poor eating habits have become the proverbial tortoise that wins the race against the faster, flashier hare. Cancer and heart disease are the leading causes of death in the United States.

Consistency is not a human trademark; all of us are risk-averse in some arenas and risk-tolerant in others. We should be most frightened of threats that have not only the greatest potential, but also the highest probability, of causing us harm. Hyperbole-free information, however, is hard to come by, and we are not always as rational as we would like to believe. Few of us don't flinch when we hear the crack of thunder, though we understand that our chances of being hit by lightning are slim. Our tolerance for certain risks may change over time. Some smokers are able to force themselves to quit only after they have children. Regan Brudie skied hard for a few winters after Todd Frankiewicz's death, but, driven in part by a feeling that he was going to kill himself if he persisted, he redirected his energies. Now if he goes skiing, it is likely to be on the bunny slopes with his children. He has never gone back to Todd's Run. Jerry Steuer, Todd and Regan's partner that day, believes that Todd's death saved many lives, including his own. The accident was a reminder that second chances don't come around very often and a call to attention. Though he didn't ski Todd's Run for ten years, Jerry hasn't backed off from high-stakes skiing. In pushing the limits, he has tried to do so wisely, with a "ski to live" rather than "ski to die" attitude. In 2003, at the age of forty-six, he let his guard down and an avalanche buried him to his neck, fracturing his scapula, lower vertebrae, and five ribs. He was back on skis as soon as his bones healed.

❄ MOST PERCEIVE skateboarding as dangerous—except, perhaps, for skateboarders themselves. The *Statistical Abstract of the United States* reports that 59,964 skateboarders were injured in 1999. But in the same year, only 3,540 fewer people required emergency room treatment for injuries inflicted by their toilets. Even more astonishing is the fact that these 56,424 people owned up to the cause of injury. Imagine the conversation behind the green curtain. "Well, doctor, my toilet has always been mild-mannered, but today it was downright vicious."

Beds are no safe havens either, claiming 455,027 victims a year. Lest you think it best to retreat to the couch, beware: an average of 330 people a day are assaulted by their sofas and davenports. Televisions are more treacherous than hammers; household containers and packaging hurt more people than ladders. Granted, more people sleep in beds and use toilets than skateboard or climb ladders. Still, you are at the gravest risk possible whenever you climb stairs or walk across a floor. It is dangerous to take ceilings and walls for granted. And don't forget to have bandages within reach the next time you slice a bagel or sit down to dinner with a fork in your hand.

Avocados rate two mentions in the *Statistical Abstract* (for production and consumption), but avalanches are conspicuously absent. Avalanches kill an average of twenty-eight people a year in the United States, roughly half the number claimed by tornadoes or lightning. Still, natural disasters—high-drama fuel for the fires of fear—are pursued so vigorously by the media that often the "storm of the century" is reported many times a decade.

In their book *Risk,* Harvard researchers David Ropeik and George Gray compiled a six-page table based mainly on 2001 data that catalogs the likelihood of a U.S. citizen being killed or harmed by an exhaustive variety of causes over the course of a year or a lifetime. To deduce the annual probability, they simply divided the population by the number of deaths from a given cause in the same year. For lifetime odds, they divided one-year odds by the average life expectancy for that year. They warn that these numbers are generalizations that do not factor in time, exposure, or demographics. The risk of being a firefighter looks especially grim for 2001 because of the toll of September 11. Obviously, you do not face 1 in 11,000 odds of drowning in the bathtub if you never take a bath. Motor vehicles rank near the top of the list, presenting 1 in 88 lifetime odds of death or injury and 1 in 6,700 odds in the span of a year. Cars kill more people than wars. In contrast, the lifetime odds of being killed or injured on a snowmachine are 1 in 94,000, with one-year odds on the high side of 1 in 7 million. Given the chance to rank dogs, deer, and bears in order of the danger they present in the United States, you are likely to place bears far ahead of docile deer. And yet, primarily by darting into the paths of oncoming automobiles, deer claim an average of 135 victims a year. Bears rarely kill more than 2. For all we love them, dogs are responsible for about 18 fatalities annually.

What is the message? Should we never sleep, favor snow-machining over walking, exercise extraordinary caution around toilets, copy the airlines and use only plastic utensils, chew our meat deliberately or become vegetarians, give away the dog, stay out of cars, and head into the snow with a sense of abandon? Those who live, work, or play in the mountains are certainly predisposed to at least the tail end of this conclusion, but of course greater exposure increases the probability of becoming a statistic. The problem is that behind every statistic is an individual with a name and a circle of relatives and friends left with holes in their hearts.

❄ FOR SOME, risk is tainted with the negative connotations of other four-letter words. But if you are taking no risks, you are dead, and without risk, we might forget we are alive. If everyone on the planet had an equally modest appetite for risk, most of our civilization's celebrated strides would never have been realized and we'd likely still be snuffling around in caves. When the Earth was thought to be flat, people worried about falling off the edge into a fathomless void. Now that we take the roundness of Earth for granted, we seem preoccupied with finding edges. Many of us live everyday lives that feel rutted enough to glorify risk.

To peruse the advertising in most major magazines is to embark on a journey of potential misadventure. When was the last time you leaped off a cliff after sipping a soft drink? Why is it that a picture of Mount Everest convinces us to buy an expensive watch or a skydiver's grin helps us choose a credit card? One automobile ad with a two-page color photograph of an SUV pulling a trailer full of glistening new snowmachines proclaims in oversize letters: THE MORE PEOPLE SCREAMING AT THE TOP OF THEIR LUNGS DOWN SNOW-COVERED SLOPES, THE MERRIER. Consequences are airbrushed from the ads as blithely as skin blemishes from the images of movie stars.

By definition, some risky ventures must fail. But as a society, we tend to be disproportionately intolerant of failure and dismissive of consequences until they shatter the crystal walls of our personal lives. An investment banker who succeeds will be praised for his daring, but if he loses your money, he is apt to be thought reckless or incompetent. A mountaineer who makes a decision, for all the right reasons, to quit short of the unclimbed summit is more likely to be ignored than hailed as a hero. If he survives and others die because they did not exercise the same good judgment, the survivor will undoubtedly be labeled lucky.

A few days before Larry Holle died in 1981, he dug a coffin-sized hole a good six feet deep outside his home in Colorado's San Juan Mountains. A friend who happened by asked why he was digging the pit. "Dammed if I know," he answered.

Thirty-eight-year-old Holle worked as a pediatrician and played just as hard as a marathon runner, hiker, and avid skier who eschewed turns for headlong downhill flings. He lived in a small building that was either a barn for a person or a cabin for a horse, the thin wall separating the horse from the man little more than a formality.

On the day he died, Larry took three visiting doctor friends ski touring in the alpine terrain of Wolf Creek Pass, turf he knew well and had skied as recently as the previous day. In the interim a foot and a half of wet, gloppy snow had dumped from the clouds. Though the men didn't observe any fresh avalanches, they chose to be conservative, skiing only on rolling, heavily forested slopes. By midafternoon they had looped back to within 250 feet of their car and were following the treed top of a ravine that dropped fifty feet to a creek, beyond which was the road. Tall, rangy Mike Davidson was trailing twenty-five yards behind Holle when he heard a *whumph.* Recognizing the sound as a sign of the snow's distress, Davidson shouted to Holle that if they ever skied this terrain again, they should invest in avalanche beacons. "Aw, Mike," answered Holle over his shoulder, "avalanches are a Fig Newton of your imagination." Those were to be Holle's last words. Davidson, following in Holle's tracks, came to a blank spot. In place of Holle's tracks was a fracture line that had shot onto the ridge from the slope below and pulled him over the edge. In his descent Holle hit a tree, broke his neck, and was likely dead in the amount of time it took to read this sentence.

Was it fate or poor choice that decided the day? Bernstein writes, "If everything is a matter of luck, risk management is a meaningless exercise. Invoking luck obscures truth, because it separates an event from its cause." The terms *hazard* and *risk* are often used as synonyms, but they are not interchangeable. *Hazard* takes into account the physical attributes of the exposure to potential danger—the speed of the oncoming car or the angle of the slope. *Risk* is the chance of something going wrong—the hazard compounded by the consequences. It's the "What's going to happen to me?" part of the equation.

Managing risk is a balancing act between a desired outcome and the probability of achieving it. Knowing your goal is key because it becomes the yardstick that helps determine how much you are willing to put at risk. Obviously, accomplishing the first ascent of Impossible Peak necessitates accepting a greater possibility of death or serious injury than going on a leisurely afternoon hike with friends. For some, taking and getting away with risks becomes the raison d'être. There is no more surefire way to feel alive than to gamble with and cheat death. Mountaineer Tom Hornbein calls these risk-seekers "adrenaline junkies." He is not being derogatory; to be forced so fully into the present and to feel as though every neuron is vibrating at attention can be addictive. My friend Jim Woodmencey, who manages risk daily in his work as a helicopter ski guide in Wyoming, believes that "thrill-seeking has become the fast-food version of adventure."

I told a friend the other day that Doug and I didn't see ourselves as risk-takers, and when she stopped laughing, I tried to explain. We spend summers rowing for months at a time in the Arctic not because we want to suspend ourselves over shockingly cold water or wiggle around icebergs the size of skyscrapers in Kevlar shells that are only an eighth-inch thick, but because we love wild country. The risks are justified by the magic of watching a polar bear nurse her cub or traveling for weeks on end encountering caribou instead of people. On the few journeys when we felt that danger and bodily torture far outweighed the rewards of continuing, we turned around.

Many remark that they'd love to undertake such trips if only they had the experience, but the reality is that to row twenty-five thousand miles, we started by rowing the first mile. A greater challenge than taking risks has been learning to minimize them—it can be harder to be cautious than to be bold. Though we have drenched and terrified ourselves in the process, we've learned to build fluidity into our decisions and now return home having had fewer close calls than on our early journeys. We travel with more food than we think we need and make sure that no one is expecting us at a given time or place. We live by our habits because they might just save us—so we always tie up the boats even when we are sure they are well above tide line and keep each other in sight even when not on speaking terms. We've found it possible to row very exposed coasts as long as we let the wind and waves dictate our moves—sometimes we've camped in one spot for weeks on end waiting for the right timing to round a tricky cape. As our technical skills have improved, we are better able to handle hazard, which also means we are more likely to put ourselves in situations where trouble is just a stroke or two away. Well aware of how thin a line we are treading, we've made a conscious decision to notch back and leave ourselves a wider margin of error. We still get to where we want to go, but now we spend the borderline days hiking instead of worrying about drowning.

Playing "could have, should have, if only" was a bad tendency of mine, particularly when I'd spent a few tedious days in the tent and was running perilously low on reading material. We work hard not to second-guess ourselves; if we've made a decision that it is unsafe to go, then we stick with it and reevaluate as conditions change. In an effort to steer clear of peer pressure, Doug and I have a standing agreement that if one of us thinks conditions are doable and the other does not, we heed the more conservative opinion.

❋ FOR ALL THE obligatory years of schooling we all undergo, we are taught little about risk management, except for whatever sex education is sandwiched into health class. Formal lesson plans rarely address decision making or problem solving. Most of the decisions people make daily are seat-of-the-pants judgments, which usually serve well enough. Survivors of avalanche accidents often say that in the seconds, minutes, or hours leading up to the avalanche, something just didn't feel right. With time to look, the reasons underpinning the emotions can usually be identified— maybe the slope was uncomfortably steep or the survivor was reacting to the sound of the snow.

Instinct may help keep us safe if we use it to back away from an edge, but if we rely on it to assure us that everything is okay, we are likely to die. Look around the animal kingdom. Birds, squirrels, Dall sheep, mountain goats, caribou, wolves, bears, deer, elk, horses, dogs, oxen, mules, and moose have all been found dead in avalanches. (The birds, for the record, have included a chickadee, a raven, and an eagle.) It is a persistent call-of-the-wild myth that animals possess a sixth sense that keeps them from getting caught in avalanches. The biggest distinction between animal and human victims is that, except for domesticated species, animals don't have relatives who call to report them missing. Occasionally, though, an alert is sounded by other means.

In March 2002, researchers attempting to track radio-collared caribou on Alaska's Kenai Peninsula detected a number of mortality signals. After tracing the stationary collars to the base of an avalanche that had run months earlier, the biologists enlisted Doug's help and returned to investigate. A large area had been trampled by commuting grizzlies, wolves, and wolverines and was circled by a swarm of equally predatory ravens and eagles. As the helicopter landed, its rotors kicked up a cloud of caribou hair. The avalanche debris was a graveyard, a thousand feet wide and twelve hundred feet long, of smashed hooves and antlers and broken bodies. Given the degree of physical trauma, Doug thought the caribou had triggered the slide high on the mountain and taken a long ride. After laboriously sorting the skulls over a period of months, the biologists determined that the avalanche had killed at least 143 animals. Several similar sites have been found, and in only two years, nearly a third of the 700-strong herd has fallen victim to avalanches.

Sometimes avalanches claim animals as collateral damage—shortly after World War II, a slide blew a tank car loaded with Karo syrup and a boxcar filled with stationery for the Rexall stores in Seattle off a section of railway line in Montana. Both cars rolled down an embankment, causing thousands of gallons of syrup to mix with tons of paper. Given adverse winter conditions, railway officials decided to wait until spring to clean up the mess. For weeks afterward sticky, paper-plastered grizzly bears could be seen feeding in the area. Boxcars full of corn have also been derailed by avalanches in the same stretch; bears gorging themselves to drunkenness on the fermenting grain have then been struck and killed by passing trains.

Doug and I have taken inspiration from animals in labeling some of the human tendencies that get people into trouble in the mountains. There is the "sheep syndrome" (blindly following whoever is leading); the "cow syndrome" (stampeding back to the barn); and the "lion syndrome" (rushing to lay down first tracks). The first time I stood at the top of an intimidating slope in the company of avalanche experts, I thought them excessively polite. Shuffling their feet, they kept offering one another first tracks, "No, it's all right, you go ahead and go first." Really, they were just astute enough to know there wasn't much percentage in being a lion.

When Doug started teaching people how to travel in the mountains, he focused on the physical parameters that make avalanches possible—the slope's angle, aspect, and anchoring capability, the layering and bonding of the snow, the recent influences of precipitation, wind, and temperature. In a late-night epiphany, he figured out that this scatter of data could be arranged in a triangle, the sides labeled *terrain,* *snowpack,* and *weather.* But though we try, most of us don't keep our thoughts neatly ordered like clothes hangers in an immaculate closet. Our minds tend to be more like vacuum-cleaner bags, with shiny pennies and paper clips sucked in amid moldering bits of food, dust balls, and clumps of hair from the dog. Even if we are trying to pay attention as we glide through the mountains, a typical bag of thoughts might look like this: *I'm ready for lunch. I can't feel that rain crust under my snowshoes anymore. Wish I hadn't eaten my chocolate bar already. It's been a little windier up here. I like Tom's friend—I wonder if he's hooked up with anyone. The snow is getting a lot deeper. I wonder if anyone else is thinking about how steep that face is. Did I mail that letter to Mom?*

The human stick figure that Doug had drawn inside the data triad was casting too long a shadow. By paying the most attention to the pieces of information that best suited their goals and needs, travelers were mimicking the Texas sharpshooter who shoots the side of a barn and then draws bull's-eyes around the bullet holes. Heeding only favorable data or making decisions by default or happenstance upped the chances of being blindsided. Groups could always fall back on majority rule, but were the resultant decisions better? As blue-collar philosopher Eric Hoffer said, "When people are free to do as they please, they usually imitate each other." What we needed was a kind of job action for thoughts, a way to get organized and prioritize data. Doug thought that a checklist approach might work—something similar to what pilots use before takeoff so that throttling a plane into the sky is not a hit-or-miss proposition. He built an avalanche hazard evaluation checklist around four main questions: *Is the terrain capable of producing an avalanche? Could the snow slide? Is the weather contributing to instability? What are your alternatives and their possible consequences?*

The checklist helps us sweep our thoughts into a tidier pile: *The slope angle on the face is 39 degrees. Looks like storm winds pillowed some snow in here. It's pretty smooth—boy, the blueberries were great here last summer. Haven't seen any avalanches. This snow feels different than at lower elevation; it's cracking around our tracks. Nice summit day. My toes are getting cold. I wonder if anyone else is ready for lunch. That ridge looks like a much gentler route than the face. But Tom's friend seems to be going for it. He's in good shape; I'm glad I'm keeping up. I wonder if he has a girlfriend?*

We've graduated from the vacuum-cleaner bag. Now we're more like answering machines. We can take messages, but we can't guarantee what portions will be heard or whether static will interfere. Don Bachman, an avalanche forecaster who speaks in a voice that will never need a microphone, has said for years that it would be convenient if snow about to avalanche would turn a telltale color. As it is, all we can do is drag traffic signals into the mountains—at least metaphorically. Assigning each critical piece of data in the checklist a colored light depending upon conditions helps lower the subjectivity in go/no-go decisions. Red lights signal, *Stop! Danger! A hazardous situation exists.* Green lights say, *Go, it's okay.* Yellow lights indicate, *Caution,* on account of potential hazard, too much uncertainty, or changing conditions.

A 39-degree slope, wind loaded and smooth, is bright red terrain. We can venture there, marry Tom's friend, have kids, and maybe even live to see grandchildren graduate from high school as long as the snowpack and, preferably, the weather are green lights. We can also ignore a red-light snowpack if we stick to green-light terrain. Most avalanche accidents happen when the terrain is a red light, the snowpack is a red light, and the skies have cleared, with sparkling green-light weather that entices people into the mountains.

So, here we are, staking lives on colors when what everyone would really like to do is add up 2 and 2 and know that it equals 4. Some argue that if travelers are already ignoring obvious physical clues, they are just as likely to allow human factors to cloud their judgments, even if made aware of the negative impact on their objectivity. In Europe rule-based decision making—a kind of cookie-cutter if-then approach—has been introduced as a strategy, particularly for newcomers to the mountains with limited avalanche skills. Ultimately, the checklist affords greater freedom and works under the broadest variety of conditions, but only if we wallpaper it to the canyons of our minds, stay vigilant, and don't go color-blind.

———

❄ A MONTH AFTER Todd Frankiewicz died, I was asked to simulate an avalanche accident as training for the local mountain rescue group. Still reeling from Todd's death, I didn't feel much like playing war games that Saturday, and my mood wasn't helped by the snow conditions on Friday afternoon. I couldn't see any way to create a scenario even close to realistic given an armor of ice on the slopes. Snow doesn't get much more stable than it was; a chain saw would have been a more appropriate tool than a shovel for building a fake debris pile. But it began snowing after dinner, and from bed I heard the first blasts of wind hit the house. By morning I needed a shovel just to exit the front door.

On Flattop, the same mountain where Nick Coltman would lose the use of his legs years later, a short slope called the Tunnel sits only a ten-minute walk from the parking lot. Steep, smooth, and loaded by southeasterly storm winds, it has the same physical attributes as the more prominent and commonly feared Death Gully around the corner. But because it is only about 150 feet high and bisected by a summer trail, it is typically overlooked as a potential hazard. Avalanches in Death Gully have caught at least twenty-six people and killed one. Avalanches in the Tunnel have caught no fewer than ninety people, including eight members of a mountaineering class practicing safety procedures. There have been no deaths yet, but there will be.

At nine on Saturday morning, I rendezvoused in the Flattop parking lot with five mountain rescue team members who would help me set up the scenario before the rest of the group arrived. Pointing our noses into the wind, we headed straight for the Tunnel. I was sure that if it had not already avalanched and produced the perfect practice location, I could easily make it do so. My partners followed me

up a safe route to the top of the slope where we linked our
arms together and, on cue, leaped into the air. We landed all
at once, as hard as we could, butt first on the snow, right
above the convexity where I had seen the slope fracture
many times. The snow stayed put. Undeterred, we moved
horizontally and jumped again. And again. And again. We
jumped ten times without result. My normally good-natured
partners were beginning to grumble. It was blowing forty
miles an hour and snowing sideways. All we'd done so far
was help snow infiltrate our hoods and trickle down the
backs of our previously warm necks. We had better things to
do with our day.

  With the Tunnel as my destination, I'd known before I'd
even risen from bed that I had red-light terrain, red-light
snowpack, and red-light weather conditions. Before we had
left the parking lot, I told the group that if we couldn't make
the slope avalanche, we would have to move somewhere
safer to set up the exercise. But, as my partners were pointing out with increasing conviction, we hadn't seen other avalanches or shooting cracks, and we hadn't heard any
whumphing noises. We had called a temporary halt to our
banzai jumps when a hiker appeared beneath us at the bottom of the Tunnel. Seemingly oblivious to both the blizzard
and the potential danger, he had marched right up the throat
of the 36-degree slope, energetically kicking steps into the
hard, drifted snow. I'd been sure that our rescue skills were
about to be tested, but the hiker had breezed past us without
incident or even a glance at our row of sitzmarks.

  Despite the consensus, the decision about whether the
Tunnel was safe enough to trust was mine to make. We had
punched it impressively hard without result. The temptation
not to waste time finding another spot was palpable. Maybe

the snow in the early part of the storm had been warm and wet enough to stick to the ice layer, creating a bond capable of holding the subsequent load. Sometimes wind-loaded snow is virtually pasted to the mountain. I could feel the undertow of peer pressure—if everyone else thought the slope was okay, then it probably was.

With the wind in my face and my goggles fogged, Todd materialized in the mist of my thoughts like a commuter elbowing his way through the crowd toward a departing train. The terrain and weather were still blazing red lights, and I hadn't seen enough to assume that the snowpack was anything but red. Maybe the only reason we weren't seeing other avalanches was that we could barely see. Maybe we just hadn't hit the sensitive spot on the slope. One easy alternative was to go somewhere perfectly safe; there is enough risk in real rescue missions to make it imperative not to jeopardize lives in practice. Still, I couldn't make myself give up on a slope I knew should avalanche. "Let's try something else," I shouted into our huddle.

Staggering in the wind, we skirted the western edge of the Tunnel and retreated downhill on tundra sandblasted bare. Triggering avalanches on foot from below is generally unwise. At best, it can result in a Darwin Award nomination and, at worst, it is career shortening. I was, however, beginning to sense that the only way to wake this snoozing beast was by stepping on its toes. Often wind slabs are thickest near the top of the slope and thinner downhill. Just as it is easier to find the pickle in a deli sandwich made with one slice of pastrami rather than ten, it can be easier to disturb the underlying weak layer where the slab is inches rather than feet thick.

With half of our contingent safely stashed as sentries, two volunteers and I began to traverse diagonally uphill in a loosely spaced line. I can hear my mother asking incredulously, "You did what?" But—in my own twist on a *but*—we were on one side of the path, on a triangle of slope only fifteen feet high. Because we were traveling in the direction of escape, toward the extreme eastern edge of the Tunnel, the sliver of threat kept narrowing. Rather than trying to skim lightly over the drifted snow, we stomped like square dancers. In the lead I could hear the weaker, drier, colder layer from the calm early part of the storm beneath the slab. Knowing that we'd found our spot, I yelled, "Get ready!" just as the snow broadcast its own warning with an inch-wide crack that shot uphill from my boots.

We had caused the weak layer to collapse beneath the slab and on top of the bed surface of ice. The bed surface is important because its surface area is roughly a hundred times greater than all of the other slab boundaries combined. (Picture yourself as a hard slab—which for most of us takes some imagination. You have the greatest bond with your bed if you lie on your back or stomach as opposed to standing on your head or turning on your side. Again, it is a surface-area issue.) When the bonds holding the slab to the bed surface give way, the stress is transferred to the top, sides, and lower edges of the slab, which can rarely hold up to such strain. We had purposely undermined the slab's compressive support, and the fracture tearing across the snow above us indicated that the tensile forces holding the slab in place had also capitulated. All was unfolding according to plan except that the three of us were still within the flanks of the avalanche when we were upended. Instead of exiting to the side like actors who have played their parts, we were on our way downhill, riding atop an unbroken island of snow.

Gravity, though, had a bit part in this particular drama. It only had the chance to split our block into a few smaller pieces and knock us askew before the snow plowed into the flats at the bottom and stopped. Brushing ourselves off with bravado, we rushed around the corner into the main part of the Tunnel to see what we had wrought. Big avalanches on small slopes are like the shy mousy cousin who suddenly bursts onto the screen as a sexy movie star. Sections of the slab were deeper than five feet, and some of the blocks were larger than full-sized refrigerators. Together, my five fit partners barely managed to lift a chunk the size of a love seat that weighed a minimum of six hundred pounds. More than twelve feet of angular hard slab debris covered the bottom of the slope. The scenario we set up looked so authentic that many of the arriving rescuers thought it was real.

What scares me still is not the avalanche but how willingly my group followed my lead and how close I came to ignoring blazing red lights. If we had quit trying to trigger the slope after our banzai jumps, the Tunnel would have either caught us as we set up the scenario or released under the weight of dozens of people running around with their beacons set to receive rather than transmit. My complacency would almost certainly have killed. And there would have been no way to rationalize that.

❄ EARLY IN MY CAREER, when my confidence was still unmarred by nicks and war wounds, I attended an international symposium of three hundred avalanche field-workers and scientists. I spent the first three days marveling that so many people had gathered from more than twenty countries just to talk about snow. I felt like a dog who had been raised among cats and then suddenly found her own kind. When a noted Indian scientist with whom I'd been having an in-depth conversation about temperature gradients pointed to a carrot on an hors d'oeuvre table and asked me what it was, it struck me that the language of snow was universal. I could go to India and be wretchedly lost in the cities yet completely at home in the snow.

The keynote speaker was André Roch, the pioneering Swiss avalanche expert whose half century of work lies at the core of modern avalanche forecasting. A wizened sprite of a man in his late seventies, Roch barely cleared the podium as he spoke in a thick French accent. Lined with details, his story took some time to tell. In essence, his son came home from college one winter break eager to ski. The snow was unstable but André figured that, as director of the Swiss Federal Institute for Snow and Avalanche Research, he of all people should be able to determine a challenging but safe place to go skiing. He described the day and the slope, his pride in his son still evident as he helped us imagine his son's first perfect turns. Then his voice faded so low that the audience leaned forward and we lost our breath with him as the snow shattered and an avalanche carried away his son. André faltered as he searched for a way to describe how it felt to have just killed his son. Begging speed from his skis, he flew down the slope to search the debris. When he said he found his son a little broken but alive, audible relief rippled through the room. André paused in his conclusion to wag a thick arthritic finger at us. "Remember," he said, "the avalanche, it does not know you are an expert."

Experts don't make expert mistakes but regular old garden-variety ones. A carpenter with thirty years' experience cuts off his finger with a table saw not because he isn't aware of the danger, but because he has become casual around it. In sawing through thousands of boards, he is also more likely to encounter the nasty knot of probability. Skills and experience are like parachutes: they can protect you from a fall or they can send you on the ride of your life. Doug took such a ride in February 1992 when he left me in Alaska's single-digit temperatures and headed for balmy California.

❄ FOR THE FIRST two days of the three-day avalanche workshop Doug was teaching at a ski area near Lake Tahoe, he felt as though he'd stumbled upon paradise. Temperatures were in the sixties, and the sun shone with an intensity Alaska wouldn't see until late spring. On the third morning, Doug wanted his group of professional patrollers to check out a fracture line produced by torrential rains the week before. The snow was now "disgustingly stable rhino hide" so avalanches weren't a concern, but the slopes were covered with a layer of glare ice. Doug's group was following the gentle treed ridge they'd decided offered the safest route to the far side of the bowl, when the two patrollers in the lead zipped off on a shortcut. "I was at a disadvantage not knowing the local terrain," says Doug, "and when I saw the greater exposure of this route, I stopped the rest of the group to discuss the good, the bad, and the ugly."

The slope below them "looked like a hockey rink tipped on end." If anyone fell on the impenetrably hard ice, he or she would find it virtually impossible to stop before hurtling into a wall of trees several hundred feet downslope. Everyone in the group "smelled the skunk." But—the area of exposure was short, and all were expert skiers who had managed this kind of treacherous terrain dozens, if not hundreds, of times. Resolving to take special care, the group took the shortcut.

It was Doug who, on new skis with unfamiliar quirks, lost his purchase and fell. In a flash he was accelerating as though flooring the gas pedal of a car, spinning as he slid. Doug kept his head up and eyes open. He knew he had little prayer of stopping, but that didn't prevent him from trying. Flailing at the slope with his poles, he managed to slow himself to a hopeful fifteen miles per hour before dropping over another edge. Now rocketing down a slope as steep as a ski jump, he saw a stout pine tree with stubby branches protruding like unwelcoming arms roughly 150 feet below him. Seconds later, traveling at least forty-five miles per hour, he slammed into the base of the tree broadside with what was later calculated as four thousand pounds of impact pressure.

Doug hit on his right side and was knocked eight feet back uphill. He says the crush of pain was "like nothing I ever want to experience again." His first thought was that he was dead; he couldn't fathom how surviving such a crash was possible. If he wasn't dead, he must have severed his spinal cord. He saw the scene unfold as though he were outside his body, perched on the limb of a nearby tree. Slumped to the right and partly curled in a ball, with his skis pointed downhill, he was making a hideous, uncontrollable moaning sound. The air seemed to suck from his lungs for a full thirty seconds. He knew that when the cry ended, he would either be dead or his breath would begin to return.

The first quick-thinking students to reach Doug removed his skis and jammed them on end like fence posts to keep him from sliding farther downhill. Within three minutes Doug found enough breath to reassure them that he wasn't about to die. Supervising his own rescue, Doug was

Doug Fesler, circa 1982, at −20 degrees Fahrenheit
ALASKA MOUNTAIN SAFETY CENTER COLLECTION

packed onto a stretcher and lowered by rope, painstakingly and painfully, down the mountain.

About the same time Doug was being x-rayed, scanned, and admitted to a California hospital, I was climbing the mountain behind our house with a friend named Jerry Bell, a mild-mannered, self-effacing emergency room doctor. My intention had been to take the day off from both exercise and avalanches. The week had gone full tilt, and I'd spent the morning digging my car and Jerry's Suburban out of deep snowdrifts. But this mountain looms large in my sense of place; it is the first thing I see when I look out the window from bed each morning. Doug and I have hiked up its flank almost every day we've been home for the last twenty years. I know where the moose like to nurse their young in the spring and where the best blueberries are to be found in the fall. In winter the mountain has been my most steadfast tutor, allowing me to observe changes in the snow day by day, even minute by minute. I know the slope angles within every path, and which are carpeted with velvety fields of heather. My favorite chair in the living room has become an observatory from which I can spy on the activities of four separate avalanche paths. So when I looked out the window at lunchtime and noticed that the path we call Ski Bowl had produced a good-sized avalanche, something it does not do every year, I felt I had to pay it a visit. Jerry, who had never seen an avalanche up close, asked if he could come along.

Only one of two starting zones in the path had released, so I chose a conservative route, favoring the side that had already slid, as we climbed the thirteen hundred vertical feet to the fracture line. I kicked steps into the hard bed surface with my bright orange plastic climbing boots. Though he didn't say anything to me, Jerry—skittish of heights and wearing much softer leather footwear—was uncomfortable on the 37-degree slope. Still feeling the strain of four straight days of avalanche teaching, I was giving him ten-cent explanations as we toured the avalanche rather than more generous descriptions of what had happened and why. If Doug had been there, no elaboration would have been needed. I stored away the details for Doug; this was the biggest avalanche I'd ever seen in this path, and I knew he would be disappointed to have missed it.

Jerry and I climbed the last few feet to the ridge and stood for a moment enjoying the spread of Anchorage below and the white sail of Denali 150 miles away. But I was fixating on the left-hand starting zone that had not yet avalanched and was tantalizingly wind loaded. We wound around until we were standing just above the rock rib that separates the two bowls. "I'm going to see if I can make the bowl go," I told Jerry, "and if it doesn't, we'll descend the same way we came up. Watch me." Only later did I discover that Jerry had no idea I was going to try to make the slope avalanche. When I spoke of making the slope "go," he assumed I was looking for a route that would be easier and less nerve-racking for him. He was watching me, but he was not at all prepared for what he was about to see.

Placing my heels carefully, I worked my way down the rocks. They were thinly covered in depth hoar that poured in around my boots like dry sand. With plenty of weak layer but no overlying slab, there was no hazard until I reached the base of the rocky spine. Although he gets in trouble every time he does so, my friend Bruce Tremper, director of the Utah Avalanche Center, likes to say, "Wind slabs are like women. They are curvy and seductively beautiful on the outside. But if you get on the wrong side of them, they can be dangerous as hell." When probing along the edge of the bowl confirmed that three feet of wind slab lay atop the same weak layer I'd just been walking through, I felt like a dowser who had found water. I moved my ski poles into self-arrest position, holding them diagonally in front of my chest as though to ward off evil. With one hand gripping the handles and the other near the baskets, I planned to use them as an emergency brake. I had four steps to make this work; to advance any farther beyond the cover of the rocks would be too dangerous. "I'm ready," I called back up to Jerry. When I asked him if he was watching, he responded with a cheerful shout and a casual parade wave.

I took two steps. *BAM!* With the report of a rifle, a chasm shot open beneath my feet and I was dropped onto my butt. From eye level, I could see snow geysering as the fissure zippered all the way around the bowl and began to widen. Below me I could see the bottom of the slab buckle. For one or two glorious seconds, I was surfing a three-foot-thick, 150-foot-wide board of snow. A gentle *swoosh,* like the sound you might hear inside a dishwasher, filled my ears. Worry didn't crowd my mind; I was at the upper edge of the avalanche with no snow that could overcome me from above.

After sliding a hundred feet, I could feel the mounting speed and turbulence of the avalanche and knew, with an ache of regret, that it was time to take my leave. I rolled onto my stomach, dug my ski pole baskets and the toes of my boots into the bed surface, fishtailed a bit, and skidded to a stop. With my head turned to the side, I watched the avalanche fall away from me, now a jumble of jostling blocks behind a mounting head of powder.

I jumped to my feet, both fists pumping the air, and watched the avalanche spend the last of its energy a thousand feet below me. As I looked back toward Jerry, my cry of triumph caught in my throat. He was shaking as though the entire ridge had turned to jelly. "Are you okay?" he shouted in an equally tremulous voice. When he asked whether it was safe to come down, I answered gleefully, "It is now!"

We compared stories all the way home. Secretly, I couldn't believe such a smart person could be so dense. How could he not have been thinking avalanche when we'd spent the afternoon looking at proof of the instability and I'd told him exactly what I was going to do? He was kind with me, but I'm sure he thought me emergency room fodder.

I pranced around the house drinking hot chocolate and admiring my handiwork through binoculars and didn't notice the bright red blinking light on our answering machine for quite a while. Spread forty-five minutes apart, two messages from the ski patrol director in California urged me to call without delay. I forced myself to sit while I dialed. Bad news isn't always something you have to be told.

The director picked up the phone on the first ring. The news came fast. Doug was alive. He was in the hospital. He was badly hurt. At least seven ribs were broken in eleven places and he had a punctured lung. He was expected to remain in the hospital a good part of the week. I dimly remember saying thank you. I have a vague recollection of punching strings of numbers, and interminable waits set to music I didn't care about. At last Doug was on the other end of the line. His every move and breath felt like a shard of glass, and coughs were excruciating, but he sounded reassuringly like himself. I slept on Doug's side of the bed that night and found myself putting on one of his shirts the next morning.

Embarrassed by his fall and accustomed to overcoming adversity through willpower, Doug pushed himself to be walking the halls by evening. He was trying to play by familiar but, in hindsight, utterly inappropriate rules. The next morning, less than twenty-four hours after he hit the tree, the medical staff agreed to release him.

Doug drove himself to the Reno airport, parked his rental car, and dragged his backpack, duffel bag, and skis to the check-in counter one at a time, thinking this a moderate-enough course of action to preclude a porter. By the time he reached the gate, he was light-headed and felt his sight going "fuzzy." Suspecting that he was going into shock, he set the timer on his wristwatch as a frame of reference. Later a witness told him that his eyes had glazed over, his face had turned chalky white, and he'd begun pouring sweat. Then his head tilted back, he began moaning, and he passed out.

When Doug came to twenty minutes later, the other passengers had boarded and he was surrounded by a small squadron of paramedics, security guards, cleaning women, and flight attendants. The paramedics kept remarking on Doug's low blood pressure, and when they asked if he was a runner, he answered proudly that he climbed a mountain virtually every day.

Doug, determined to get home, talked his way onto the plane. Vomiting often but surreptitiously, he gritted through two airport changes and a four-hour layover in Seattle. At last he shuffled stiffly into my arms in Anchorage, and we went home to be together. For the next five days, we felt the strength of our luck, though sometimes Doug hurt so much that he thought the lucky ones were those who died. Getting Doug out of bed took the stuffing out of both of us, and walking upstairs left him without any breath to take for granted.

Though Doug still couldn't put on his own socks, in another self-diagnosis he thought exercise would do him good

and insisted that we go for a walk. He needed thirteen minutes (timed, again, by his stopwatch) to reach a ramshackle
cabin he normally breezes past in four minutes, and he returned winded, pale, and spent. I reported this to Jerry Bell
when he called, as he had all week, and he was quick to say
that something didn't sound right. On his way home from
work a few hours later, Jerry walked through our door,
thumped on Doug's back, put a stethoscope to Doug's chest,
and told us to get in the car and not stop until we reached
the hospital. More X-rays confirmed what Jerry already
knew. Doug's right lung had collapsed, probably the week
before when Doug had struggled into the Reno airport and
lost consciousness. Fully a third of Doug's blood supply had
seeped into the lung and congealed into a hard glob. By
morning Doug was undergoing surgery, which, despite grim
odds, saved his lung.

Jerry undoubtedly left our house incredulous that two
presumably smart people could be so ignorant about something as elemental as the human body. We'd sat home for
days ignoring clues. The irony of our role reversal was not
lost on me. My smugness of the week before peeled from me
like a shabby overcoat, and I have tried to leave it crumpled
in a dark corner.

While Doug was recovering in the intensive care unit, a
newspaper reporter wrapping up an article about avalanches
kept our answering machine busy. When he finally tracked
me down in the hospital, I told him that no article about the
Alaska Avalanche School would be complete without comment from its founder, who was currently bumping along
inside a morphine cloud. I knew the reporter well enough to
trust him, and after offering to help him sort the lucid bits
from the nonsense, I handed the phone to Doug. Though

Doug would intermittently nod into sleep in the middle of a sentence, he was remarkably coherent.

We were still in the hospital when the article ran and hadn't had a chance to read it when the reporter called, apoplectic with apology. An editor had tampered with his prose. The passage he was concerned about read: "'There is absolutely no reason in the world why you can't ski the steepest chute or gully: it is just a matter of good judgment and timing,' says Fesler, who is recovering from a near-fatal fall down a steep chute." Though laughing was agonizingly painful, Doug couldn't stop. Rarely has a newspaper article ever gotten it so right.

CHAPTER 7

Silver Screens

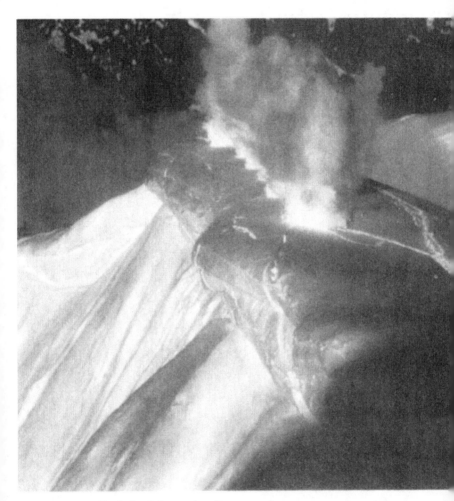

Explosives being used to cleave a cornice
© CHUCK O'LEARY COLLECTION

One is tempted to say that old myths never die; they just become embedded in textbooks.
　—Thomas Bailey, U.S. historian

Thanks to the movies, gunfire has always sounded unreal to me, even when being fired at.
　—Peter Ustinov, British actor

Doug looked and smelled more feral than the wolverine. Both had gleaming eyes, emitted the thick stink of musk, and were splattered blood-red. With a grunt, Doug heaved the wolverine over the edge of the ridge. We held our breath as it fell, first skidding along the snow and then cartwheeling over cliffs. It was only a limp dark dot when it slid to a stop near the valley floor.

"Cut, it's a take." After two straight days of throwing the stuffed ketchup-smeared wolverine off the top of the mountain and jumping into a helicopter to retrieve it, we at last had the shot the director had framed in his mind. We could have been filming a Johnson & Johnson commercial lauding the toughness of their dental floss, which I'd used to stitch an avalanche beacon inside the wolverine's rear end. Against the double odds of rough treatment and my poor sewing, the mint-flavored floss had held. But this wasn't an ad, it was an action movie called *Running Free*. Our services are not required for artfully conceived classics. We work on two kinds of films, at opposite ends of the cinematic spectrum: documentaries and action-adventure flicks that tend to make only cameo appearances on the big screen before being relegated to video stores. In service of the latter, we blow up cabins, fake air crashes, suspend screaming women over rivers raging through rocky gorges, drop young boys into glacial crevasses deep enough to swallow a ten-story building, and set off avalanches.

Avalanches in the movies have a tradition of being almost better than the real thing. They can last several minutes and are set to soul-stirring music. When James Bond is caught in a monster that looks like it will best him, he has only to push a button and he is cocooned within a giant silver balloon from which he emerges with his smile and his hairstyle intact. Things are looking grim for the backwoods boys in the classic western *Seven Brides for Seven Brothers* after they snatch six potential wives from a frontier town in Oregon Territory circa 1850. The posse is gaining on their horse-drawn wagon, but the brothers have a plan as they hurry through a steep snow-filled canyon. They know one of the most basic movie myths—that loud noises trigger avalanches. The eldest brother claims to have cussed his horse so vehemently one hard February that the resulting avalanches kept his clan snowed in until spring. So the handsome brothers carefully muzzle the hysterical maidens with their work-hardened hands until they reach the upper end of the canyon. Then, with a cry of "All right, let 'em go," the girls are allowed to shriek while the boys unholster their revolvers and shoot volleys willy-nilly into the air. The mountain cooperatively unleashes a series of avalanches that seals the canyon with such a wall of snow that the posse is forced to admit defeat and rein their horses back to town, giving the boys a few isolated months to win the girls over.

If only real avalanches had movie endings. The wickedly

handsome downhill ski racer in an old Rock Hudson movie eludes the avalanche licking at his heels by vaulting into a tree, and rescuers commonly find the missing with a few shallow swipes of their hands. CPR works with one breath, leaving the victim with enough air for a passionate kiss. As British director John Boorman has said, "Movies are the repository of myth." Given that Doug and I have spent careers debunking misperceptions about avalanches, when we received a phone call one October in the early 1990s asking if we could spend a day in March making avalanches for a feature film, it was not without a sense of irony that we said yes. Thus was launched a sideline business of creating "designer" avalanches.

If you've ever seen contemporary footage of an avalanche rushing headlong down a mountain—its cottony cloud filling more and more of the screen until, as the symphony on the sound track cranks up to fortissimo, there is nothing but white and the avalanche pushes you into the back of your chair and sweeps over your head—chances are good that we threw the explosives that triggered it. Actually, you are likely watching at least half a dozen different avalanches spliced together in a kind of avalanche greatest hits. If you pay attention, you will over time come to recognize individual segments like reliable old friends. We were in the movie theater watching Brad Pitt in *Seven Years in Tibet* when one of our own avalanches surprised us by charging onto the screen.

The plot—though that is a generous attribution—of the first movie we worked on called for the old standby rifle shot to send millions of pounds of snow into motion. A loud noise is as unlikely to trigger an avalanche as I am to break a champagne glass by singing. When Doug and I showed reluctance to perpetuate this myth, the director agreed to a change in script. The mad trapper (actually Doug dressed as his stunt double in a fur suit and bushy hat) would hang out the open door of a helicopter shooting at a wolverine he was determined to see dead. But with each miss, his wrath would mount until, seething, he'd draw back a dirty canvas tarp and snatch a handful of dynamite conveniently stashed in the rear. We were filming, with me pressed into the backseat where the cameras couldn't catch me pulling the fuses and handing the sputtering charges to Doug, who was almost blind beneath the hat, when I smelled the unmistakable odor of burning fur. With seventy-five pounds of explosives at our feet, it wasn't a good time for surprises. Cameras be damned, I patted down Doug as thoroughly as I could and urged him to be expedient in heaving the remaining charges out the door.

Guaranteeing an avalanche six months in advance is a trick. But it helps to buy a generous pile of explosives and pick out some fat, well-overhung cornices as understudies. If need be, the cornices can be enticed to drop. As the hard blocks careen down the slope below, they will scoop up the loose surface snow and deliver a convincing avalanche, complete with powder cloud. One moderate-sized cornice—say, fifteen feet wide, fifteen feet thick, and fifty feet long—can tip the scales at more than 360,000 pounds and break into chunks heavy enough to dig furrows four feet deep into the snow on their descent. Doug describes it as the equivalent of rolling tractor trailers filled with cement down the slope.

Our ideal candidate is a slab just unstable enough that it will crumple to pieces if we bomb the right spot, but not so hair-trigger that tiptoeing into the bottom of the path to set up a camera is too risky. The need for the slope to be perfectly illuminated further narrows the field. A cloud in the wrong place can keep us sitting on a ridge for hours and may force the day's shooting to be scrubbed. We could be performing surgery, given the split-second timing required.

Often we work with Steve Kroschel, a self-made avalanche cinematographer and feature film director who is so smitten with avalanches that one always makes him ravenous for another. Steve, who made his first movie about a day in the life of a muskrat, says, "I've filmed a lot of different things in my life. But the avalanche is so powerful. It's almost a spiritual experience for me. When we capture these images on film, it's forever. It's permanent, and it's shared with millions of people." Steve's feature films may have a more limited audience, but much of his footage is used in documentaries produced by National Geographic, Discovery Channel, and others, and so is seen globally.

Steve multitasks as scriptwriter, animal trainer, cameraman, fund-raiser, and lunch-maker. Boyishly good-looking even in his thirties, he still has the body of the Junior Olympic weightlifting champion he once was. He is a kind of idiot genius who will ask three times in rapid succession whether it is legal for Doug and me to have different last names though we are married, and then will fleetly invent a remote-control camera-triggering device from a rat trap and a $3 egg timer as a substitute for fickle high-priced models.

Typically, while I sit on a ridge, duct-taping two-foot-long orange plastic tubes packed with high explosives into bundles and feeding in lengths of fuse with blasting caps crimped on one end, Doug and Steve tackle more creative logistics. They use the helicopter to lower a reinforced steel "crash box" into the bottom of the path and strap two avalanche rescue beacons to the outside, wrapped in foam and sealed in Tupperware. Then they carefully insert a movie camera with a price tag greater than our collective net worth. Steve takes extra time to make sure the lens commands the kind of view a person would never want to have, the equivalent of lying in front of a speeding freight train.

Mounted on a scrap piece of board, the egg timer will trip the rat trap, which in turn toggles the switch on the camera. Once Doug and Steve set the timer and latch the top of the steel box, we have exactly fifteen minutes before the camera begins to roll film. They jump aboard the helicopter and hustle back to the ridge, where, with the rotors still turning, I load several boxfuls of bombs and climb into the rear seat beside Doug. The door next to us has been removed, and we are loath to rely on seat belts alone, so, cinched into climbing harnesses, Doug and I clip into safety lines secured to the central metal framework of the helicopter. I wear two watches—one to track the time on the rat-trap remote and the other to monitor the burning fuses—and a belt pack with pull-wire igniters, crimpers, and other tools of the trade. Nothing can sit loose on the seat or it will be sucked into the sky or, worse, the rotors. We have already done several dry runs with the pilot, practicing the elevation and speed with which we want to move along the ridge and selecting our targets. Ideally, we crab sideways no more than ten to fifteen feet above the snow so that we can more precisely position the charges and so that my taped bundles are less likely to break apart upon impact.

We plan to throw three or four charges that must be lit all at once so that they will explode more or less simultaneously. This means we have live charges at our feet for the half minute it takes to scoot along the ridge between the first bomb drop and the last. *Fifteen minutes on the egg timer, two minutes of film in the camera, ninety-second fuses on the explosives.* If we move too quickly, the action will be over before the camera switch is tripped. If we are too slow, the camera will run out of film before the avalanche occurs. As we circle, I cut a half inch off the top of each fuse to ensure that the tip is dry, reducing the chances of a dud, and seat the pull-wire igniters firmly on top of the fuses. The igniters look like long, thin cigars but act more like firecrackers. When the string on top is pulled, a little burst of flame shoots out the bottom of the cardboard cylinder and sparks the fuse.

I do most of the talking because Doug is leaning far out the door, and much of what he says into the voice-activated microphone on his headset will be lost in the rush of wind. If he needs to tell me something to relay to the pilot, he signals with his head or left hand. *Thirteen minutes thirty seconds on the rat trap. Get ready.* Doug balances the charges between his knees and grasps the tops of the igniters in one hand. The pilot lines up the chopper on one end of the run. *Fifteen seconds, ten seconds, five seconds, pull.* I sniff acrid smoke and look for telltale heat bubbles in the plastic coating of each fuse; we have *fire in the hole.* "Fire in the hole," repeats the pilot, acknowledging that he has live bombs behind his back. *Fourteen minutes on the rat trap. Ten seconds on the fuses.* Doug points toward the mountain and then, with his palm flat, motions as though he is patting the head of a dog. *A little farther right, slow down.* Doug heaves the charge with both hands, as if he is slinging a bag of flour. *One out. Two out. Fourteen-ten on the rat trap, twenty seconds on the fuses. Lower.* Doug's forefinger twirls. *A little faster. Three out. Four out. Okay, all clear.* It is cold, really cold—with the door open, the temperature with windchill can drop to more than 20 degrees below zero. Our fingers are numb as the pilot swings away from the ridge, and we back off to watch the show, careful not to taint the sunny slope with the chopper's shadow. With twenty seconds left on the fuses, I begin another countdown for the cameramen we've positioned safely on various ridges.

We can't hear the charges explode from within the helicopter, but we count the tall plumes of yellow smoke to make sure we have no duds. If a bomb doesn't detonate and doesn't fall with the avalanche, we'll have to wait thirty minutes and then return to the ridge. The pilot will hold a low hover while one of us belays the other out onto the thin skid to place a live charge on top of the dud.

When all goes well, several hundred feet of cantilevered cornice cleaves between black craters, leans as if in slow motion, and then thuds onto the underlying slope. For a moment the blocks just bounce at awkward angles—sometimes completely perpendicular to the fall line—but then the avalanche begins to gather itself, springing forward in muscular leaps like a tiger. I can almost hear the sound track in my head.

Our strategy is to position the crash box high enough in the path that it will get dusted by the avalanche, but not so far upslope that the camera is likely to be obliterated or deeply buried. If we see no sign of the box as we fly over the debris, we land, switch our beacons to receive, and fan out to pick up the signals. Steve usually runs alongside us like an exuberant puppy. Finding the box, however, doesn't mean that the camera is intact or that it rolled film. A debacle during the shooting of *Running Free,* in fact, along with a few equally demoralizing miscarriages, spurred Steve to design the rat-trap remote.

That day, the double beeps my beacon was receiving grew stronger as I climbed to the right. The same signals were pulling Doug left from the far side of the debris. When we converged, I handed my probe to Steve, who jabbed the snow with abandon that would have been alarming if we were looking for a person. The probe struck metal with an unequivocal *thunk,* and I swung the shovel off my back, knowing that we had the box.

Even before the box was chipped free, Steve was down on his knees, scraping snow off the lid with his bare hands as if he'd discovered the buried treasure he'd been hunting for a lifetime. He pried the lid open, reached gingerly through the foam, extracted the camera as if it were a day-old baby, cuddled it to his chest, and let out a moan that would have triggered the whole mountainside if sound could do such a thing. The camera had slept through the whole glorious avalanche. At $700 an hour for the helicopter, not to mention a host of other costs, it was an expensive snooze.

The rat-trap remote has yet to let us down. Steve, though, is still prone to grabbing our hands and praying fervently that the camera rolled film before he dares open the lid of the box.

❄ MAKING MOVIES has ruined the way we watch movies. Nothing is as it appears, and rarely is a sequence filmed in order from beginning to end. The more dramatic or dangerous the scene, the smaller the pieces it must be broken into to be filmed. Describing one scene he wanted us to set up, Steve said, "I want to have the boy walking high on a *Sound of Music*–type mountain, and then slip and fall a thousand feet down a really, really steep glacier before he catches himself with his fingernails on the lip of a huge scary crevasse, and then two ravens start to peck on his hands, and he starts sliding, and he's kicking and barely hanging on when One Paw [the star wolverine] grabs his sweatshirt and saves him." My mind was still trying to catch up with the action when Steve, like a kid overcome on Christmas morning, began asking, "Can you do that for me? Can you do that? Do you think you can do that?"

If the boy takes the fall on a steep glacier full of holes, he will die. Instead, as the boy ambles across a snowy ridge, he passes a hidden pit where Doug and a friend lurk out of view of the camera. They yank his ankles so that he falls abruptly. *Cut.* On a steep but short slope, with nothing nasty at the bottom, we let the boy slide and tumble to his heart's content while his somersaults are filmed. *Cut.* With unseen slings looped around the ankles of both the boy and the cameraman, who are positioned nearly side by side, we run downhill, pulling them as fast as we can so that the fall looks out of control and bumpy. As the boy fights to arrest his fall, the cameraman is at his elbow recording the struggle. *Cut.* With our shovels, we build a four-foot-high snow slope above the edge of a carefully selected crevasse. Our "hill" will be spliced in so that it looks like the bottom of the slope where the boy took his fall. The boy wears a harness and a wetsuit under his clothes, and we tie him in to our anchors with white ropes hidden underneath the snow. We've engineered everything except for a convincing look on the boy's face. He doesn't appear nearly as scared as he should be while careening toward a bottomless-looking hole in the ice. Of course, he knows that he will drop only so far before the rope stops him.

We are now two months into a movie for which we had originally been hired for a single day. A helicopter has been picking us up at the top of our driveway every morning and returning us ten hours later. Our valley has a row of rural mailboxes clustered where the paved road degenerates to dirt. One night another resident notices the address on my box and asks if I know anything about the helicopter landing twice a day on Brewster's Drive. I am counting the minutes until I can go to sleep and play dumb for fear of a diatribe about noise. "Well," says the woman in a tone that implies more fact than rumor, "I hear those people are running drugs."

On the glacier our patience is eroding. We have a long list of scenes to shoot, and in take after take, the boy's expression of fear looks faked. He is cold and whiny, and I have to keep warming his fingers on my stomach. Doug takes aside the boy's mother, who has been scrutinizing our every move, and asks if he can build an extra foot of slack into the belay system without telling the boy. For some reason she says yes, and when the boy flies past his usual stopping point without being braked by the rope, a perfect look of unadulterated terror flashes across his face. *Cut.* Part of the hanging-on scene is filmed from below by a cameraman we have lowered into the cold crevasse, but the flailing feet in soccer shoes and the wriggling butt are actually mine. I am secretly thrilled that my butt can double for that of a twelve-year-old boy, though the likeness is admittedly helped by distance. *Cut.* Even the ravens don't want to go near the boy because he is making such a fuss, but we throw some worms close enough that the cameras can grab a point-of-view shot establishing the boy and the ravens in the same general vicinity. *Cut.* When it is time for the ravens to peck, Steve offers up his own hands. *Cut.*

We have been willing to feed raw chicken to the two wolverines (both males, but mates for the Disneyesque purposes of the movie) and take them for walks, but the nose-to-nose encounters are Steve's to stage. Whenever possible, we use the wolverine named Skippy as a stunt double for One Paw, so named for his one white front paw. For the closer shots, Steve slops a little paint onto the ever-cooperative Skippy. When I ask Steve why he thinks Skippy has a mellower personality than One Paw, he answers that his wife breast-fed Skippy alongside their son. Doug, horrified, quips that perhaps we should call Steve's wife "One Tit," but the discussion is clearly over. When it comes time to film One Paw clawing his way out of the mad trapper's avalanche, he keeps popping effortlessly from the snow. We enlarge the burial hole we've dug in the hard slab debris but flatly refuse to crawl into it with a half-tame, bad-tempered, fully-toothed wolverine. Instead, Steve is soon beneath the snow hanging on to One Paw's back legs in an effort to make the wolverine's escape look harder. *Cut.*

❄ STEVE'S AVALANCHE ambitions know no bounds. For a while he was content to hit snowmachines, derelict buildings, and blow-up dummies. But by his third movie, *Escape from Alaska,* he wanted a helicopter to be parked in the path of a hard-charging avalanche. Envisioning the aspects of the scene that involved the movie's hero and heroine was easy enough. They would be filmed sitting in a helicopter grounded in a parking lot, with plenty of time to shout at each other, grimace, and jiggle the joystick. When it came time to film the scary part, though, it would be our sorry butts on the line. If there was a pilot fool enough to wait until a full-blown powder cloud with hurricane-force winds was almost upon him and then cold-start his helicopter, we didn't want to fly with him. Instead, we scouted and scouted from the air until we found a huge bowl funneling into a canyon that might give us our avalanche, Steve his shot, and the helicopter an acceptable margin of safety. A second helicopter with cameramen aboard would jockey slightly higher and a few hundred feet behind us, filming our attempt to escape the avalanche. Added to the usual list of hazards to be juggled was the very real threat of a midair collision.

When the stakes are high, we practice. Only when we were confident that all players understood their parts did we leave the ground for more rehearsals. The vertical drop of the path was almost 4,000 feet—for comparison, if the World Trade Center's Twin Towers had been stacked one on top of the other, the vertical drop of the resulting 220-story structure would have been 2,730 feet. If we were able to spawn the avalanche we wanted, it would fly off a cliff about 2,000 feet into its run and channel into a tight-waisted rock gorge. Feeling the constriction, the avalanche would climb the walls of the canyon, shooting two or three hundred feet into the sky. We'd be lying in wait, and when the powder blast shot out the bottom end of the gorge, we planned to rise just above it.

Steve wanted to film the avalanche from the helicopter's front passenger seat, and we agreed, on condition that in the heat of the moment, the only person the pilot would take direction from was Doug. We dropped the charges and did a roller-coaster dive toward the valley bottom to get into position. As we hovered nervously, the cliff constricted our view. We weren't even sure we'd triggered an avalanche until a cameraman on a nearby ridge gave a yelp over the radio to let us know it was coming. And suddenly the avalanche was cascading over the cliff edge like Niagara Falls. As it grew in his viewfinder, Steve became more and more excited and started directing the pilot, "Go down, go down, go down!" The avalanche exploded up the canyon walls; Doug and I could see that the powder cloud was going to rise even higher than three hundred feet. It had the white billowy look of a cumulus cloud. "Go up, go up, GO UP NOW!" Doug commanded the pilot in a tone that brooked no debate.

The avalanche passed less than a hundred feet beneath the belly of the helicopter. Out the open door, we could see boulders of snow flying, colliding, and spinning within the powder cloud. Images of heat came to mind as I peered into the boiling, bubbling, cauldronlike heart of the avalanche. Doug keyed the mike and in Alistair Cooke's *Masterpiece Theatre* voice said, "But it seems to have a greater gravity than a mere vapor. It descends too steadily and too compactly to be only a fitful fog." The pilot probably thought Doug was cracking under pressure, but I knew he was quoting from an unpublished manuscript by Neil D. Benedict, one of his "bonanza" finds while researching Alaska's avalanche history. Benedict, part of the Gold Rush throng, had seen "an American snowslide . . . carrying enormous trees on its grinding, crushing, roaring bosom" as the wooden sloop carrying him north from Seattle limped into Valdez Bay in April 1898. With wide eyes and a purple pen, Benedict concluded, "How you strain your eyes to take in every detail of the marvelous spectacle as you begin to have some dim notion of the force by which worlds are hammered into shape from the molten members of an infinite Cosmos."

Every bit as enthralled as Benedict, we had initially hesitated at the prospect of accepting payment for the privilege of triggering avalanches purely for their beauty. We shelved our scruples without second thought, however, when a contractor who hadn't had to risk his hide hanging from helicopters or keeping company with bombs charged $6,000 to throw wet salt through the windows of a dollhouse while the action was filmed in slow motion. Call us stooges—we'll admit to being scared as his "avalanche" burst into the house and filled the living room with "snow." Our own house, reproduced by Steve as a scale model, makes a brief but unpaid appearance in the same movie when it gets hit by an avalanche and explodes into flame.

In feature films our own on-screen appearances are generally limited to select body parts, though Doug still receives an average of $4.65 (pretax) in annual royalties for his "acting" part in *Escape from Alaska*. Naturally enough, he plays the lead rescuer, who must call off the search for the heroine because half of Juneau "has been buried . . . and with a couple of hundred people missing, we still have a chance to save some lives." But the voice that emerges from his lips is startlingly alien—his meager three lines have been dubbed. Of course, the hero carries on with his search and rescues the heroine on his own. Realism may not be Hollywood's strong suit, but lines like "She can't be dead!" or "We're talking about a ten-year multimillion-dollar project, and you're telling me about clouds!" ring all too familiar.

The line between fantasy and reality has a way of blurring when filming avalanches. During the shooting of a National Geographic program in 1995, we loaded the crew and all their gear onto snowmobiles and rode several miles into Chugach State Park. The survivor of an avalanche that had released the previous winter came with us to tell his story in the shadow of a large bowl that had killed his friend. "We were just riding," he said as the camera rolled film, his eyes wide as an owl's. "I had to go to work at six. We just wanted to come out and play a little bit because it was the end of the year."

When he finished, Doug and I took up the rescue strand of the story. When the avalanche occurred, I was elsewhere in Chugach State Park, having taken the day off to ski with my visiting parents. I'd been nervous enough about the snow to leave my pager on my hip and my rescue pack in the car. When the page came, I drove quickly to the parking lot that was serving as a staging area, gave my parents a kiss and the car keys, and told them not to wait up.

With pilot Bob Larson, Doug had been flying home from the Kenai Peninsula, where he'd just dug a snowmachiner from under fifteen feet of debris. Over Turnagain Arm, Bob remarked how delighted he was that he was going to make it home in time for dinner. Laughing, Doug replied, "I wouldn't count on it, with the snowpack the way it is." Less than ten minutes from town, the radio crackled with news of the most recent accident, and Bob turned the helicopter into the mountains. The fracture line that awaited us there looked uncannily the twin of the one that had killed another snowmachiner in the same path four years earlier.

When the interviews were over, the producer wanted to film a rescue scenario. We took out our shovels and had clues scattered about and people buried in the snow when one of the rangers we were with received a call over his radio looking for us. Several houses had just been hit, and homeowners in the area were worried about the possibility of additional avalanches. Doug and I abandoned the fake rescue and within fifteen minutes were in a helicopter whisking north to Eagle River.

Another time, for an action scene in one of Steve's movies, we loaded an airplane fuselage onto a trailer, hooked it behind Doug's Toyota pickup truck, hauled it up the frozen Matanuska River, where we set up a high-line system to winch it onto the snout of the Matanuska Glacier, and lowered the wreck into a crevasse so that it hung as though it had just crashed. Months later Doug was summoned to Canada, where a film crew had pushed the limits in remote mountains. This time, instead of being hired to keep them safe, Doug's charge was to retrieve their bodies as well as pieces of the helicopter from the glacier they had augured into.

❄ KEEPING AS MANY as fifty actors and crew members from triggering avalanches, plunging into crevasses, walking into helicopter rotors, falling off ridges, or freezing body parts can be a rodeo—especially when they are unfamiliar with the environment, preoccupied with their own objectives, and hurried by time and budget constraints. In a scene that could be a comedy, a documentary crew once filmed us scouting locations with Steve Kroschel. He, of course, was looking for perfect lighting and spectacular scenery while we needed triggerable avalanches and safe-enough terrain to be able to position the crash box. As we flew over hanging blocks of ice the size of warehouses and crevasses big enough to hide cathedrals, Steve kept repeating, "This is a good spot, isn't it? Doesn't that look good to you?"

During the filming of a commercial, Doug once had twenty Eskimo dancers, twenty-five dogs, four dog mushers, and a dozen Japanese cameramen and actors high on a glacier, with a storm rapidly approaching. Doug made sure that his rules were explained—in Yup'ik, English, or Japanese, as appropriate—to each of the players. He never deduced what the commercial was about (cigarettes were his best guess, based on the number consumed), though he did manage to get everyone flown back to town as night and a full-blown blizzard closed in.

Sometimes we feel as though we may as well be speaking Yup'ik for all the attention a crew is paying to us. One spring the cameraman for a public television documentary about sea kayaking in Prince William Sound decided that he wanted the host of the program and Doug, who was doubling as the on-camera "talent," to sit in their kayaks beneath a towering blue iceberg that was grounded on the beach. Doug refused, not only because he thought the iceberg's collapse was imminent, but also for the bad example it would set. Ignoring Doug, the cameraman had his helpers drag the boats into position as he began trying to scale the slope of the iceberg. We exhausted every polite form of explanation until finally Doug exploded, "I want both boats out of there NOW, and I won't take no for an answer." Throwing in a few expletives for punctuation, he had just dragged the second boat clear when the iceberg splintered and crashed so dramatically that it would have flattened the kayaks and killed or badly injured anyone below. We didn't have any trouble being heard from then on.

Doug and I are very conscious that we are most likely to make high-consequence mistakes if we make other peoples' time pressures or expectations our own. We carry in our minds the example of a filmmaker who ventured into Colorado's Berthoud Pass in 1957 eager to attain unparalleled avalanche footage for a Walt Disney movie. He achieved his goal—but with an outcome he hadn't anticipated. He was manning his camera on the road while highway crews used a 75-mm howitzer to shoot the twin starting zones of a major path known as the Dam Slide. He had been warned that avalanches could reach the road but judged his position a reasonable risk since this had not happened for twenty-four years. The first two shots hit their marks but did not entice any snow to slide. The third round sent both bowls into motion at once, generating a much larger avalanche than anticipated. Through the camera's viewfinder, the avalanche would have looked perfect, the powder cloud flicking eighty-foot-tall trees into its maw as though they were matchsticks. Only when more than 120,000 tons of snow rushed the road did the photographer comprehend his jeopardy. As his camera was knocked askew into the snow, it captured a split-second glimpse of the highway worker who had been standing near him running full bore down the road. The last frame of the footage is shockingly white.

Bulldozers were used to find the photographer. Even after his body had been extricated, the sixteen-foot wall of snow that had killed him was imprinted with the haunting figure of a fleeing man.

CHAPTER 8

Line of Fire

*Avalanche at Bird Flats that blocked the Seward Highway,
damaged the power line, and threw a section of railroad tracks
into the water, 1988* © DOUG FESLER

It is the state policy that emergencies are held to a minimum and are rarely found to exist.
    —by Alaska law, section 44.62.270, State Policy

The fact is if avalanches do nothing else they teach humility.
    —Monty Atwater, *The Avalanche Hunters*

AFTER THE JANUARY 26, 2000, AVALANCHE IN CORDOVA that put Jerry LeMaster in the hospital, it continued to storm. The sky might as well have opened bomb bay doors and torpedoed snow—seventy inches fell in less than a week in the heaviest-hit parts of southcentral Alaska. By late Monday, January 31, more than a dozen avalanches were blocking the Seward Highway between Anchorage and Girdwood. When an avalanche nearly waylaid his truck, Terry Onslow, the seasoned forecaster in charge of the Seward Highway's avalanche program, wisely withdrew his troops and prohibited road crews from attempting to clear the slides. "I don't know if I saw the avalanche move or not," he told an *Anchorage Daily News* reporter. "You couldn't see more than fifty feet." Raised in the harsh semi-arid plains of central Montana, Onslow rations his words like water and allows himself a wry smile when he describes an especially large slide as having plugged "five lanes on a two-lane highway."

Novelist Julia Alvarez has described the disorienting swirl of whiteout as "not just snow but snow in a tantrum, snow angry at being used for too many pretty winter scenes in postcards and poems, snow proving it can be mean and serious." As he drove north toward Anchorage from his home on the Kenai Peninsula that Monday, Darwin Peterson was in the teeth of the meanest blizzard he'd experienced in his seventy years. The road and sky were an undifferentiated smear, and winds of over fifty miles per hour were shaking the truck as though it were prey. Before Peterson could reach Girdwood, a large avalanche slammed the highway in front of him. He and three other drivers pulled U-turns, but their convoy managed only a quarter of a mile before another slide buried the two lead vehicles to their windows, pinching their escape route shut. Now seven adults and two seven-year-old girls were stranded between the two slides like rats in a trap.

Peterson called for help on his cell phone. The Alaska State Troopers immediately arranged to have a rotary plow from the south extricate the group, but yet another avalanche inundated the roadway. With shredded timber mixed into the debris this time, the available blower was rendered powerless. Rescue from the north was impossible because of the slides from which Onslow had already retreated. Not even the bravest or most foolhardy rescue pilot could fly a helicopter through the blizzard, so Peterson's group was relegated to spending a long night on the road. With little gas to spare, they could only fire up their engines for several minutes each hour in a failing effort to keep the chill at bay.

Two decades earlier, on a gorgeous March day, Doug had come upon thirty vehicles and a road grader stymied by an avalanche that had just blocked a quarter mile of Seward Highway. In the deceptively benign weather, a jovial atmosphere had taken hold, and the drivers were standing outside their cars chatting and clicking snapshots. One look at the slide, however, filled Doug with urgency. It was big, bigger than any he had seen in that spot—so impressively huge that he later dubbed the path "Superman's Climax," a name that has stuck. "We have to get everyone out of here now," he barked, enlisting the grader operator's help. A scant half hour after they had accomplished their task, the sunny pavement where the crowd had gathered lay beneath twenty feet of snow debris from a second avalanche.

Avalanches are like fish: they tend to run in schools. When one has occurred, more are likely. Ideally, the highway would have been gated closed before Darwin Peterson became trapped. This would have involved running sweep, which means driving to wherever the road is blocked, quickly trying to ascertain whether anyone is caught, turning stopped traffic around, and making sure all vehicles are clear of the avalanche area before barring the road behind them. It is a scary race against time and consequences. Often the drivers idling at the bottom of paths that have not yet run are oblivious to the danger. But if you are the person running sweep, you know that this is not the time for discussing how long the road might be closed or what the weather has in store for tomorrow. You direct the drivers to turn and take your place at the back of the crawling procession. Drumming the steering wheel with your hand, you mutter a mantra of "Go, go, go!" as you flash wary glances uphill, wondering if you are about to be blown off the road. Any veteran of sweeps has stories of white clouds materializing in the rearview mirror or slamming on the brakes and skidding sideways as the leading edge of a slide swirls before the headlights.

Unable to drive north to Anchorage or south to the Kenai Peninsula, motorists luckier than Peterson were stacking up in Girdwood, where the storm had forced even the ski resort to close. With snow plowed as high as the street signs, the antennas on pickup trucks looked like periscopes and the roads might as well have been tunnels. And still it snowed. Some motorists found shelter on the floor of the school gym while others surrendered their credit cards to hotels, moved in with friends, or curled up in their vehicles. There was a touch of festivity, a "What can you do?" mood that would degenerate when it became clear that the party was going to last well into the week.

On Tuesday morning, February 1, there was a break between storms. It wasn't more than a gasp—a temporary lifting of the clouds and a reprieve in the winds, which were forecast to redouble to a hundred miles per hour by afternoon—but it was enough to allow a swarm of helicopters into the sky. With both the Seward Highway and the Alaska Railroad knocked out of commission, the airspace along Turnagain Arm became a whir of activity. The Air National Guard was one of the first to launch, sending a Pave Hawk helicopter from Anchorage to evacuate Peterson's group of nine from their cars. Terry Onslow climbed aboard an Alaska State Trooper helicopter to survey the slides, while other avalanche workers with the Department of Transportation and the Alaska Railroad crammed a bigger ship with explosives. Doug—on special assignment to the State Emergency Coordination Center to assess the avalanche hazard to a vulnerable neighborhood and determine whether it was safe to open the school in Whittier—strapped himself into the front seat of still another helicopter. In Girdwood a group of six had been stuck since Super Bowl Sunday when in slinky dress clothes they had taken a limousine for a birthday celebration. Now they were each gladly shelling out $127 to charter a helicopter for the fifteen-minute puddle jump to Anchorage. "Hallelujah, guys, dinner is finally over!" exulted one woman, unaware that a stampede of others, some cradling pet turtles and dogs, would soon copy this last-resort method of exodus. The Troopers also authorized use of the Seward Highway as an airstrip so that small fixed-wing planes could come in for their share of the commercial kill.

In copies of the *Anchorage Daily News* dropped at doorsteps that Tuesday morning, I offered the cheerless prediction: "I'd be very surprised if we get through this cycle without damaging a house or killing more people." Not an especially difficult call, it would be proven right in hours. With bad roads, shrieking winds, zero visibility, and Herculean trail breaking, most backcountry aficionados would elect to stay home. Instead, I was more worried about the punishment avalanches would inflict on power lines, transportation corridors, and exposed neighborhoods. I knew that the fleecy hip-deep powder I'd waded through in my driveway that morning would be whipped into spring-loaded slabs as soon as the wind picked up. Even if these slabs weren't going to add stress to a tender weak layer that wasn't even marginally bonded to a bed of ice, there would have been a fusillade of avalanches. But with these layers in place and warming temperatures and intense snow-transporting winds, the setup couldn't have been more classic—or more deadly.

I was at the squat no-nonsense operations building of Chugach Electric Association by eight. Doug and I had signed on as the company's avalanche safety consultants in the late 1980s, though it had taken me a few winters to earn the trust of those I was supposed to protect. At first, even when I spoke, the responses were directed to Doug. His gender gained him easier entry into a fraternity of canvas overalls and dented aluminum lunch boxes, and it didn't hurt that he was older and had more experience. Plus he looked the part—Popeye to my Olive Oyl. But the credibility gap was long gone by the time I stood in Chugach Electric's headquarters that morning, my feet sweltering in the same white bulbous rubber "bunny boots" favored by the linemen.

As I snatched breakfast from a desktop candy dish, the operations chief sketched the problem as he knew it. Sometime in the night, the high-voltage power line between Anchorage and Girdwood had been breached. Most likely it was a casualty of an avalanche at Bird Flats, the first major clutch of avalanche paths twenty-five miles south of Anchorage. No one had yet been able to get close enough to locate or size up the damage, so there was still hope that the towers were standing and only the conductors were down. If so, it might be possible to effect a quick patch.

For the moment the company was feeding the stranded population of Girdwood electricity from the "back side"— that is, by diverting power north from generators on the Kenai Peninsula. Additional avalanches, however, were certain in the next twenty-four hours. If any of these slides damaged the southern end of the electrical grid before the link at Bird Flats could be reestablished, Girdwood as well as a host of other communities would be cast into the dark and cold indefinitely. As if there weren't already enough helicopters in the sky, the operations chief, crew foreman, and I commandeered one of our own. Their job was to assess what it would take to resurrect the power line; mine was to determine whether it was safe to work.

The boxy buildings and concrete plain of Anchorage slid from view in minutes, replaced by the majesty of Turnagain Arm, the cul-de-sac of ocean named by a presumably disgruntled Captain Cook during his unsuccessful search for the Northwest Passage in 1778. (With tides of nearly forty feet, Turnagain Arm boasts one of the largest tidal ranges in the world. Cook's ship was at the mercy of currents that washed him back and forth like a Ping-Pong ball every six hours, forcing him to "turn again" more times than he wished.) As we rounded the bulky shoulder of Penguin Ridge, Bird Flats came into sight, the parallel strands of power line, road, and railroad looking like little more than offerings at the base of mile-long slopes that rear four thousand feet from sea to sky.

The avalanche chutes along Bird Flats are officially numbered one through seven, but most are known by more descriptive names such as the Dogleg, Straight Shot, and Five Fingers. My first thought was that for all the fuss, the offending slides looked puny. I'd been anticipating thugs like the ones in 1988 that had ripped seven transmission towers from their foundations, flung the wires into the ocean, and heaped the highway and railroad tracks with forty-foot mounds of debris. Now, though an expanse of road was buried, the debris from the avalanches looked to be a measly fifteen feet high. But volume wasn't the only measure of their power. Reid Bahnson, a hard-bitten forecaster who manages avalanche risk for the Department of Transportation north of the Arctic Circle and helps out when needed along the Seward Highway, remembers, "Yeah, those 2000 slides were Bruce Lee avalanches. Bruce Lee might have been a small fighter, but no one in the world ever hit harder."

What had given these avalanches their punch was powder

blast, the advance guard of displaced air and suspended snow particles leading the turbulent charge down the mountain. At the southern end of Bird Flats, in a path called Sneaky Pete's, a transmission tower built of three wooden poles two feet in circumference and fifty feet high had been reduced to stumps by powder blast that must have been pushing 150 miles per hour. Three strands of 1.25-inch conductor, each with a breaking force of twenty-seven thousand pounds, had been frayed into hairs and uselessly kinked like Slinkys. Even before we'd made our first aerial circle, it was obvious that there would be no quick fix.

My decision was easy. The repair crews needed not hours but days, during which they'd be sitting ducks at the bottom of Sneaky Pete's. Even if I cleaned out the intricate network of narrow gullies still bulging with snow above the work site by dropping explosives from a helicopter, if the weather forecasts were correct, the wind would reload the gullies within a few hours. Toggling the microphone switch on my headset, I explained over the intercom that it was pointless to do anything until the weather granted us a longer window of opportunity.

❋ AS OUR HELICOPTER banked toward Anchorage, we crossed shadows with the larger helicopter chartered jointly by the railroad and the highway department, which was gaining altitude to do a bombing run. The three forecasters inside it were friends and mentors of mine. Together, they had a century of frontline avalanche experience. To manage risk for ski resorts, ski guiding services, and transportation corridors is to walk a tightrope between making decisions that are too conservative and those that are not conservative enough, while juggling expectations, economic realities, and safety. If a forecaster closes a road or favorite powder run every time there is a hint of danger, he will not only lose credibility; he will lose his job. A wrong decision in the other direction could lose lives. "Excuses are never in order when you do serious avalanche work," says Reid Bahnson, who was one of the crew in the helicopter. "You make decisions and live or die by them."

Avalanche control is the presumptuous label given to the imperfect process of trying to keep avalanches from killing people or damaging property. *Avalanche reduction* is a humbler term for the same effort. There are three basic choices: avoid, defend, or attack. Avoiding exposure altogether is the most reliable long-term mitigation option, but this is not always feasible or even desirable—ski areas would quickly lose popularity if they closed off all steep and deep terrain. Constructing structural defenses can lessen the impact of some avalanches. Attacking—shock loading the snow at times when it is vulnerable—is done with the intent of inducing relatively frequent small avalanches. The danger of allowing the snowpack to build without regularly shedding layers is that any avalanches that do occur are likely to be larger and to pack a greater destructive wallop. Foresters attempting to forestall rampaging superhot wildfires are advocating the same basic strategy when they argue for frequent controlled burns.

The premeditated release of avalanches was first used not to protect people or property, but as a method of attack. During the three gloomy winters of World War I, upward of forty to sixty thousand men were killed by avalanches. Some were deliberately triggered, though certainly nature was the most powerful antagonist. Mathias Zdarsky, an Austrian avalanche expert who trained alpine troops, observed, "The

mountains in winter are more dangerous than the Italians."
On December 12 and 13, 1916 (thereafter known as "Black
Thursday" and "Friday the 13th"), six thousand soldiers died
in avalanches in just forty-eight hours. On February 28 of
the same year, Zdarsky was called to the site of a barracks
"built in an obvious avalanche path." Releasing on its own
volition, an avalanche had torn through the building, killing
twenty-five Austrian soldiers at rest. As Zdarsky scouted for
a safer place to rebuild, "the sunny day turned dark." Swept
into the seething cloud of a second avalanche, he was battered by the frozen corpses from the ruined barracks as well
as by the tumbling snow. In a book published in 1929, he describes his brutal ride, which resulted in eighty bone fractures: "I could sense all the anatomical changes in me until I
reached the point where my sacrum was about to be broken
off. The pressure became even greater, my mouth had an ice
plug, my eyes were popping out, my blood was seeping
under my skin, and I felt like my intestines would be ex-
pelled. I . . . only wished myself a rapid transit to the Other
Side . . . But the avalanche slowed its run, the pressure de-
creased, my ribs cracked as though a piano was playing a
tune on my spine."

Offensive use of avalanches has not fallen out of vogue.
In the 1980s, guerrillas in Afghanistan were known to have
purposefully triggered avalanches onto Soviet troops, and
avalanches have also likely been used as weapons in the en-
during border conflict between Pakistan and India.

The more socially acceptable notion of intentionally
triggering avalanches to render an area safer harks to the
Middle Ages. In 1438 a Spanish knight-errant traveling
through the Alps instructed his party to fire their guns in an
attempt to dislodge the snow menacing their route. Though

this has become the most popular means of starting avalanches in the movies, in the real world it probably just depleted the group's ammunition. By 1933, though, the director of a mountain railway in the Alps had begun experimenting with rockets and mortars. Subsequently, the use of surplus military artillery such as the 106-mm howitzer has become commonplace, particularly in the United States, though the dwindling supply of ammunition for these long-obsolete weapons poses a looming crisis. Inhibited by post–World War II armament restrictions, Japan is one of the few avalanche-prone countries that has not come to rely upon explosives as a primary tool for managing avalanche hazard.

In the United States, the use of artillery and hand charges as a means of avalanche control was pioneered by Monty Atwater. At the close of World War II, fresh from the U.S. Army's Tenth Mountain Division—where he had seen explosives successfully put to use to trigger slides in the Alps—Atwater arrived in Alta, Utah, an abandoned silver mining camp that was turning its attention toward the exploitation of an equally precious resource, powder snow. Intending to ski at the fledgling resort and renew his writing career, Atwater instead became a Forest Service snow ranger unafraid of seat-of-the-pants experimentation. He drilled holes in cornices and tamped them with dynamite, suspended charges just above the snow surface, and lowered large bombs over the edge of ridges—all methods still in use today. Desperate for firepower, Atwater first used an 1898-vintage French cannon he commandeered from the lawn of the state capitol building in Salt Lake City.

Early rising patrollers now routinely toss hand charges onto steep slopes that could threaten skiers. Every year resorts in the United States alone burn through more than

150,000 pounds of explosives—commonly gelatin dynamite or TNT—with a fuse and blasting cap attached. Necessity has spawned a number of ingenious delivery methods such as a "bomb tram" that transports explosives to the target area via a continuous cable looped through a series of large pulleys. In my favorite homespun economical model, a patroller sitting on a rusty bicycle mounted on a platform high in the trees lights the fuse and energetically pedals the charge into place.

A friend in Colorado once lobbed a two-pound hand charge into a gully he suspected might be unstable. Instantly, he was rewarded with radiating cracks as the slab tore loose and hurtled downhill. Ninety seconds later—because the explosive was armed with a ninety-second fuse—there was a poof of smoke in the debris far down the mountain. It had taken nothing more than the weight of the charge itself to set off the avalanche; he could have thrown the orange from his lunch and achieved the same result.

Conversely, a contractor working on the Alaska oil pipeline north of the Arctic Circle one spring distrusted an expert's opinion that the slope shadowing his men was stable enough to pose no threat. Determined to conduct his own stability test, he arranged for a staggering seventeen hundred pounds of explosives to be ferried uphill and laboriously dug into the slope. Detonated simultaneously, the charges unleashed such puffy plumes of smoke that it looked as though a herd of poodles was scampering across the mountain. When the ocher smoke cleared, though pock-marked with jagged craters, the snow remained plastered to the slope. Stability is relative to the force exerted upon the snowpack. If the contractor had used three thousand pounds of explosives, he might have gotten his avalanche. The exercise, however, would have been purely academic—

the snow was so fantastically strong that it would not have released on its own.

The most unambiguous control result, of course, is for a slope to avalanche. It is much harder to know what to do when a slope you think should slide doesn't. Perhaps the shock load wasn't put in quite the right spot, or the force wasn't sufficient, or the assault was made too early or too late. As Norm Wilson, who teamed with Atwater in using explosives to secure the slopes for the 1960 Winter Olympics at California's Squaw Valley, says, "Avalanche control is a matter of putting the right charge in the right place at the right time." If explosives are used too early in the storm, before there is a critical load, or if too much time has elapsed after the storm, or if cooling temperatures are allowed to strengthen a snowpack through refreezing, the window of opportunity may be lost. Some types of instability are less sensitive to explosives than others. Infrequently, but just often enough to give forecasters nightmares, progressive weakening caused by explosive detonation can cause a slope to avalanche hours after the ski area or exposed road has been reopened.

As the sky above the Seward Highway thickened with helicopters that Tuesday morning, Joe Perkins, commissioner of the state Department of Transportation, was on hand to see what he could do to speed the opening of the road. Normally, five thousand cars and freight trucks transit the gauntlet between Anchorage and Girdwood every winter day. Two days of closure had effectively rendered the Kenai Peninsula an island and transformed the highway into a parking lot. Perkins invited the media to Bird Flats, where, according to the next day's *Anchorage Daily News*, "he hoped to put on a victory-at-hand demonstration at 2 P.M."

Road closures are anathema to most highway officials, whose abiding purpose is to keep traffic—and revenues—flowing. For every winter day the Little Cottonwood Canyon road leading to the Utah ski resorts of Alta and Snowbird is closed, almost $2 million in revenue is lost. During a hair-raising avalanche cycle in 1978, a bureaucrat denied a request to do artillery control along the Seward Highway. "There is too much snow on the road already," he told the beleaguered avalanche forecaster, "and it is poor judgment for you to dump more." Not uncommonly, the same avalanche that inspires respect among field-workers is viewed by officials as a burdensome inconvenience, even a personal affront and an embarrassment. A highway spokesman once insisted that an avalanche blocking a road was not large enough to be called an avalanche but was just a "very, very large slide." He might as well have been a team manager contending that the football star's leg was not broken but very, very fractured.

The team in the large helicopter filled with explosives was putting into play a plan they had used successfully for decades. They'd bombard the avalanche paths above the blocked highway until satisfied that it was safe to send bull-dozer crews in to begin clearing the road. With so many avalanches down between Bird Point and Girdwood and a tight weather window, there was little chance of getting the road open enough to ease the pressure of the backed-up traffic, but any amount of progress would quicken the eventual reopening.

More than a hundred starting zones loom above the mile of highway at Bird Flats, and many are subject to cross loading by wind. The trajectory from the gun mount at the northern end of Bird Flats is so oblique that deep crannies of snow remain hidden behind rock ribs. So, for reasons of both logistics and speed, helicopter bombing was the railroad and highway department forecasters' method of choice that morning. One forecaster sat next to the pilot, guiding him to the crucial spots. In the rear the explosives handler prepared the charges, and when the igniter on the fuse was pulled and there was "fire in the hole," the bombardier dropped the charge out the open door. Contouring steadily up the mountain, the crew peppered the high-, middle-, and low-elevation starting zones at Bird Flats with twenty large charges—all without provoking a response. Continuing south to other paths, they threw a bevy of bombs, only tickling loose one avalanche.

One of the major threats at Bird Flats is a large complicated path called Five Fingers, named by Doug in the mid-1970s to reflect the five deep grooves that shape the upper starting zones. All five gullies funnel into one runout zone. In the avalanche atlas that Doug and I prepared for Chugach Electric Association in 1991, we described its characteristics:

> Five Fingers . . . has the capability of producing large destructive avalanches on a regular basis . . . Debris depths from major slides have been estimated in excess of 40 feet . . . Powder blast heights from large slides can exceed 200 feet and avalanche velocities within the 80–120 mph range are not uncommon . . . This path poses a serious potential risk to personnel working in the runout zone along the power line or highway right-of-ways.

I landed back at Chugach Electric's headquarters before noon and, while driving home, saw the first few banners of snow trailing off the crests of the ridges, silent warnings that the wind was beginning to stir. A message left on our answering machine by Terry Onslow around 11:15 let us know that explosive control hadn't produced much. "Everything that wanted to release seems to have already let loose," said Onslow, adding that the "iron"—heavy equipment—was being sent in.

Larry Bushnell, John Rajek, and Kerry Brookman, who had waited patiently inside the cabs of their idling Caterpillars during the bombing, rumbled around the roadblock, out of the safe zone, and toward the toe of Five Fingers Path. Though the machines were big, they looked like Tonka Toys as they crawled over the snow debris. With the highway department's permission, a crew from KTUU Channel 2 television, the local NBC affiliate, crept in behind the operators to get good footage.

As Bill Mowl, a district supervisor for the highway department, told the *Anchorage Daily News* later that day, "When we started, it was dead calm. An hour later, it went to hell." About 12:30 P.M., John Rajek saw the cameraman running toward KTUU's vehicle and wondered, *What the hell is going on?* At the same instant, his radio squawked a frantic warning from the spotter assigned to watch the mountain: "Avalanche! Avalanche! Avalanche!" Putting years of safety training to use, Rajek, Bushnell, and Brookman swiveled their machines toward the mountain and raised the blades—a David-against-Goliath attempt to parry the anticipated blow. Inside KTUU's car, reporter Laura Papetti also saw the approaching avalanche. "It looked like everything was in slow motion and then it hit the car," she told the newspaper.

Parked near the edge of the slide, Papetti weathered the onslaught without harm. The KTUU cameraman walked away unscathed, as did John Rajek. Dressed in his bright orange Department of Transportation coveralls and black watch cap, Rajek looked like an escaped prisoner, which essentially he was. When the snow dust cleared, Papetti began waving "to the guys in the Cats to see if everybody was OK," she explained beneath huge headlines the next day. "I thought they were waving back." It took her a moment to realize that no, they were not all okay. Larry Bushnell was partly buried and signaling for help.

Inside the cab of Bushnell's bulldozer, the windows had begun to crackle and pop under the pressure of the powder blast. With his seat belt on, Bushnell could do little when the cab began filling with avalanche debris. He leaned away and tried to take the force of the pounding on his right side. "At one point, I couldn't breathe," he told the *Daily News* reporter. "I was gasping for air. I thought I was a goner. It didn't move the dozer, but I felt like I was being ripped out of it."

Running to Bushnell, Papetti found him immobilized in snow to his chest. "He was calm," she informed the newspaper. "He told me to be careful because there was broken glass everywhere." With nothing to dig with but her hands, Papetti began scratching at the snow, which was "solid as a rock." There was a lot to worry about. Could she get Bushnell free? What had happened to the third driver? Could another avalanche barrel down on them? Unlike the equipment operators, she was not wearing an avalanche beacon. Together, Bushnell and Papetti picked away at the snow, but without a shovel their efforts were futile. Bushnell found his radio and alerted coworkers that he was alive and needed more help.

Fifty-three-year-old Kerry Brookman, working about fifty feet from Bushnell, took the brunt of the hit. The avalanche kicked his 35,000-pound D6 Cat three hundred feet out into Turnagain Arm as though the bulldozer were a beer can, flipping it end over end across the frozen mudflats and crushing the cab. The powder blast continued another five hundred feet beyond the destroyed machine.

Dave Hamre, the forecaster in charge of avalanche safety for the Alaska Railroad, and Reid Bahnson were on the Anchorage side of Bird Flats helping to refuel the helicopter that had just been used for bombing when word of the slide blared over the radio. Hamre frantically pressed the pilot to hurry. Doing a "cold liftoff" in a large helicopter is like making a rock fly, but in moments the men were airborne, swooping low around the short corner to Bird Flats.

They spotted Brookman's twisted, overturned bulldozer at once and touched down a hundred feet away. As Bahnson sprinted across the frozen tidal flat toward the Cat, he saw what appeared to be a "headless body on top of the snow." It was Brookman's snowsuit turned inside out with the radio microphone and ripped headset cord still attached. Thinking Brookman was buried beneath the snow, Bahnson and Hamre were switching their avalanche beacons from transmit to receive mode when they spotted wiggling fingers. Brookman was sitting as though in a recliner, with his head under a few inches of spindrift from the powder blast and his legs buried about sixteen inches. The avalanche had stripped him down to his T-shirt and pants. Hamre and Bahnson quickly uncovered Brookman's head. Brookman was cold and in shock, but he assured his rescuers that he would be okay. As Hamre carefully shoveled Brookman free, Bahnson conducted a more thorough medical survey. He found no obvious fractures or bleeding, but through a four-inch-long laceration over Brookman's kidney, he could see obvious internal injuries. By now breathless reinforcements were arriving from the road. Improvising a stretcher from a blanket, rescuers thrust Brookman into the backseat of the helicopter, and the pilot made a beeline for the nearest hospital, twenty-three miles away.

All efforts to unplug the road were abandoned indefinitely. "I'm an engineer, with some geology training," Joe Perkins told me years later, "but I had no idea what avalanches could do until that day. To see those shredded power poles, with no snow debris around them, was unbelievable. And, even with two full careers in the military and with the highway department, I had no conception of the constant burden on those avalanche control guys—their job is as high stress as I've seen. The avalanche impressed me with the importance of giving greater consideration to avalanche-prone areas when designing and building roads."

✳ I LEARNED OF the avalanche fifteen minutes after it released, around 12:45, when the first puffs of wind were beginning to spill off the tops of the mountains and hit our house. Wind is able to deposit snow ten times faster than snow can fall from the sky. It can alter the character and decrease the stability of snow in minutes, adding weight to leeward slopes and encouraging slab formation. On one cloudless day, I watched forty-mile-per-hour winds whisk four inches of soft powder into two-foot-thick, Styrofoam-hard, hair-trigger slabs in a mere hour and a half. Less than an hour of wind had been enough to whip up the Five Fingers slide.

These were exactly the conditions that Steve Casimiro, a freelance writer from California, had on his wish list when he contacted us early in the winter. On assignment for a national magazine, he wanted to shadow us during an avalanche rampage and write about the intricacies of managing avalanche hazard. I had sent him a quick e-mail advising him to stay tuned after drenching rain in December resulted in what was dubbed the "Christmas crust." Hated by skiers, the crust was viewed with suspicion by those trained to wonder how well future snowfalls might bond to it. If the next storm had come in warm and brought wet new snow, trouble might have been averted and Casimiro might have stayed home. Instead, storms in early January dropped several feet of cold powder, delighting recreationists but worrying enough for me to dash off another e-mail to Casimiro. Then the snow stopped, temperatures dipped into the single digits, and one clear day began to string into the next. Weak, glittery faceted snow with the consistency of sugar can form in any layer of the snowpack with a strong temperature gradient, and cold weather helps create these gradients. With each day the several feet of fluff sandwiched between the cold air and the ice crust was becoming weaker. In the middle of January, it began snowing again and kept snowing in earnest—on and off, but mostly on—for the rest of the month.

Snow, which covers a fifth of the earth throughout the year, is many things—protection for small animals, the summer water supply for lawns, a nuisance to some, a thrill and not infrequent menace to others. But to snowpacks perched on steep slopes, new snow means added stress. On January 27, the day after the Cordova avalanche, I e-mailed, "If there is ever a time you should be here, it is right now." Four days later I wrote Casimiro again, "You really should be here NOW! We haven't had a whopper cycle like this for a number of years." On the morning of February 1, while railroad and highway forecasters bombed Bird Flats and Kerry Brookman and the other operators waited inside their Cats, Casimiro hopped a flight north. By the time Casimiro had reached Seattle, Brookman was in the hospital being readied for surgery.

At an elevation of fifteen hundred feet, our house is just below the tree line. The anemometer on our roof had blown apart in the last storm, but after years of living in preposterously windy mountains above Anchorage, we are confident of our estimates. When our windows begin to flex noticeably, rendering our reflections rail thin one minute and laughably bloated the next, the gusts are hitting 60 miles per hour. When they reach 75 miles per hour, toy boats can sail the whitecaps in our toilets, but at 100 the boats are grounded—the vent pipe suctions the water from the bowls. Gusts of this strength have occasionally sucked out a window—the wood frame as well as the glass—leading us to drill in three-inch screws at close intervals around the frames. Slammed by gusts of 120 miles per hour, the house groans as though it is being tortured. Under such conditions, our neighbor's house lost its roof twice in three years. Sleep only comes with earplugs, and even then we sometimes retreat to the inner hallway, far from any windows. At 130 miles per hour, the highest winds we have clocked, we begin to think seriously about abandoning ship.

By late Tuesday afternoon, when Casimiro called us from the Anchorage airport, our toilets were bone dry. Doug instructed Casimiro not to attempt to drive his rental car beyond the school bus turnaround at the end of the paved road about a mile and a half down-valley from our house. Only a high-clearance four-wheel-drive vehicle with chains would have a prayer of busting through the drifts, which were growing deeper by the minute, and only a driver who knew the road like Braille might manage to stay out of the ditch in the stinging whiteout. "I'll meet you at the turnaround. Come quickly and dress before you leave the airport," Doug warned Casimiro. "You'll need snow boots, gaiters, wind pants, a parka, and don't forget your goggles," he said, before hanging up and beginning to pull on similar attire.

The rendezvous at the turnaround went as planned except that the television crew Doug had agreed to meet for a sound bite about the siege was late. He grew increasingly impatient, knowing that the longer he waited, the less chance he had of making it home. Just as he was giving up on the crew, they arrived, shocked by the ferociousness of the weather compared to lower-elevation Anchorage. This is a familiar problem. Sometimes as I descend the hill toward the grocery store, people regard my snow-plastered car as though it is a spaceship. My parents will call from the East Coast, where winds of over forty miles an hour are a major news story, while our house is being pounded by weeks of hurricane-force winds that attract little media attention. This time the element of surprise worked to Doug's advantage. Dressed in newsroom clothes and street shoes, the cameraman and reporter kept the interview mercifully short. Still, it was dark by the time Doug and Casimiro turned for home.

During a similar blizzard, Doug and I once decided in the middle of the night that we had better move one car from our house to the turnaround so that we could be more sure of getting out to a scheduled workshop. Doug drove the truck and I followed in the Subaru, but unable to distinguish the edge of the road, I drove into a snowbank. I had just finished digging out all four wheels when Doug returned to yank me out with a towrope. As we stood shouting to be heard, I told Doug that I'd try again, but that I had never had such trouble. It wasn't a matter of seeing poorly; I couldn't see at all. He looked through the tiny opening left by my cinched hood and with admirable restraint said, "Jill, wipe your glasses."

Trying to return to the house with Casimiro, Doug made several running charges at a particularly resilient drift on the last steep hill. Each time he penetrated only a few extra useless inches, backing down blindly after each failed attempt. Eventually, predictably enough, the Suburban ended up skewed in the ditch, four hundred feet from the house. "That's as far as we're going to get," Doug told Casimiro. "We have to walk from here."

Both Doug and I are as familiar with this hill as we are with each other. With the wind at my back and my body a sail, I have flown more than twenty feet down it, never making contact with the ground. Bucking uphill against the wind typically involves groveling on all fours. Once, the only way I could manage to keep hold of a new rubber garbage can I was stubbornly attempting to tote home at an inopportune time was to crawl all the way uphill inside of it, an inelegant human snail. A dead spruce tree that stood halfway up the hill used to serve as another unofficial wind gauge—we'd measure the strength of the wind by how long we had to hang on to the gnarled trunk to avoid being blown back down the hill. The tree eventually succumbed to a particularly relentless windstorm and finished its life as firewood.

Casimiro, however, had not expected to land on another planet. Here is his story:

> Staggering with my bags in the knee-deep snow, we tucked our heads into the wind and tried to plow forward, but it was like pushing against a breaking wave. Doug attempted to hoist my ski bag . . . but was blown back in a heap. I gained just three feet before my glasses were

plastered and I was totally out of breath. Each step felt like moving under water, only with the oxygen content of high altitude. Suddenly, no matter how hard I tried, I couldn't breathe. It wasn't that I was out of breath, it was that there seemed no air to breathe. I yelled for Doug, but the only way he could hear me was if I shouted an inch from his ear.

"It's the wind!" he screamed, with hands cupped around my ear. "It's blowing over a hundred and the pressure won't let you exhale! Just breathe out hard and take your time."

Even with fearful, forceful breaths my lungs seemed to have the volume of a thimble. Doug pulled me down into a drift, where we lay atop the skis to keep them from blowing away. "It's not far," he yelled. Crawling, scrabbling on hands and knees to expose as little of ourselves as possible, we pressed forward, three feet and rest, three feet and rest. Then, like a port in a storm, there was the house, the garage, a respite from the wind, and sweet air freely drawn. "Aw, yeah," said Doug proudly, stamping the snow from his boots, "it's like that all the time."

Doug peeled off his wet clothes and whipped together his standby pasta with shrimp and Cajun seasoning. The three of us were eating by candlelight when the phone rang. Kerry Brookman was dead.

The roar of the wind and the creaking protests of the walls around us more than filled the silences that followed. As the wave of excitement and action I'd been riding for days crashed, I felt the awful vulnerability that comes with making decisions about the safety of others. So what if I'd made the correct decision that day, or fifty or a hundred or even a thousand times before? How often could I make the calls and not get one wrong? As much as experience is touted, with it comes an accumulating burden. In a voice lower than normal, Doug said, "This death thing is gonna drive me out of this field. I'm just getting tired of people dying." Without another word, he pushed his chair back from the table, cleared his plate, climbed upstairs, and went back to work.

Our up-for-anything dog took one look outside the next morning and retreated to his red plaid bed. The trees were flagellating themselves and a drift filled our front door, leaving only a thin gap at the top of the frame. We were going to have to hack our way out, and I wondered if Casimiro was cursing us yet. In an unusual twist, winds up to 120 miles per hour were not only raking the mountains but also sweeping to sea level and traumatizing downtown Anchorage, ripping off roofs, felling trees, and damaging a 747 cargo jet parked at the airport. We could go nowhere until the storm broke.

In a rare move, the State Emergency Coordination Center was called into twenty-four-hour action to help the various stranded communities cope with escalating concerns. The center had been amped up in anticipation of Y2K but, as I heard one overwhelmed official mutter during a break in an interagency conference call, "Y2K couldn't do what Mother Nature has done to us." The coastal town of Seward was running short of food and medicine, and it needed relief for the lone doctor who had been working around the clock. Girdwood, low on gas, was limiting retail purchases to three gallons a day to those with generators; the rest was reserved for road-clearing equipment and emergency vehicles. The supply of diesel, which was used to heat emergency shelters, was a worry too. Chugach Electric warned Hope's community leader that the town's power might be out for several weeks. The *Anchorage Daily News* reported other more frivolous anxieties. The manager of the Burger King in Homer said, "We're holding our breath. We're running low on buns. If we don't get them, we might have to close for the day." Troubled by the brewing frustration of travelers still stranded in Girdwood, the Alaska State Troopers attempted to close liquor stores but caved in the face of public outcry.

With multiple power lines down, several power companies had us on standby, and we had been hired by the State Emergency Coordination Center to apprise them of the danger to communities and individual homes. In the video footage shot by Casimiro as he moved between Doug's office and mine that frenetic Wednesday, both phone lines are in continual use, and as we talk, Doug and I are in motion—pulling out maps, taking notes, scrolling through weather information downloaded from the Internet, petting the dog. I tell a reporter, "This is not a day for traveling around. It is just a day for being smart . . . This storm has been too much, too fast." Doug warns the State Emergency Coordination Center, "When you have unusual avalanches caused by unusual conditions, you need to exercise unusual precaution."

When Chugach Electric called with an update that afternoon, we knew we could rule out extracurricular activities for the foreseeable future. An avalanche more than a mile wide had just annihilated another seven transmission towers near Lower Summit Lake, south of Girdwood. Now much of the Kenai Peninsula was without power as well as access, and Girdwood was in the dark. Once we could get there, it would be up to us to assess the hazard, bomb the slopes if necessary (we would end up triggering slabs ten feet deep), and take responsibility for the safety of the crews over the weeks of repairs, telling them where and when they could work. Avalanche forecasting is more a preoccupation than an occupation. You think of the snow as a living, breathing creature and try to stay one step ahead of its moods. Your mind is sifting through the snowpack's stratigraphy while you wash dishes. Waking in the middle of the night, you are as likely to get up and check remote weather stations on the computer as you are to roll over and go back to sleep. If crews are working around the clock, you may well keep vigil at the site. With a toothbrush, a sleeping bag, any food you can scrounge, a cell phone, skis, binoculars, and your dog in the car, you might be away from home for days, snatching sleep in naps. A cold, dry, stable snowpack might start avalanching like crazy after only a half hour of light rain. A slope that is frozen and stable in the morning might be trouble as it warms and weakens in the afternoon. Mark Twain wrote, "There's something fascinating about science. One gets such wholesale returns of conjecture out of such a trifling investment of fact." In avalanche forecasting, you rarely have all the facts you want. To stay on the winning side, you have to keep shuffling the deck and examining the cards in your hand.

Our phone began ringing in the middle of the night from faraway media either oblivious of or unconcerned with the time difference. At 3:30 A.M. Alaska time, the Weather Channel in Atlanta, Georgia, wanted confirmation of their theory that the avalanches were being caused by El Niño, the warm Pacific current. More than one reporter from a big-name newspaper did not want to hear that similarly big avalanche cycles had happened before and would happen again, and refused to accept the fact that global warming had nothing to do with the current chaos. It is not a new phenomenon to feel a need to assign blame in the face of a crisis.

When the weather finally relented enough on Thursday for us to escape the house, the three of us spent most of the morning sitting in the Suburban at Bird Flats, parked at the overlook that now bears a bronzed memorial to Kerry Brookman. Next to us was a large truck loaded with fifty-pound bags of ammonium nitrate, a relatively cheap and readily available explosive that is simply a mix of diesel fuel and fertilizer. Defiant but spooked, a little army of avalanche workers was escalating the war. Our answer to an instability that had proven so stubborn and dangerous was a lot of bombs—no less than eight thousand pounds of explosives—so many that, for a while, the supply was in question. Highway workers had inventively fashioned a gun sled out of an old pickup truck bed, with pieces of guardrail for skids. At low tide, around noon, the workers used a bulldozer to pull a howitzer mounted on this sled out onto the ice-covered mudflats of Turnagain Arm, far beyond Kerry Brookman's ruined Cat and hopefully beyond range of any descending avalanches. From there they had a much more direct shot line into the starting zones. They fired twenty-two artillery rounds, with virtually no results. When the clouds withdrew sufficiently, a chartered helicopter lifted off and bombs rained from the sky. Doug climbed aboard and dropped even more charges in the path at the far end of Bird Flats where Chugach Electric crews needed to erect a new tower. The mountain above Bird Flats became so riddled with holes that there was hardly a virgin bomb placement to be found by the time any of us were willing to accept that the snow had surrendered and was stable enough to trust. Larry Bushnell and John Rajek were among the drivers who lined their bulldozers up end to end on the barricaded road, waiting to attack the debris.

The Millennium Avalanche Cycle, as the avalanches of late January and early February 2000 became known, took two lives, buried miles of highway, and hit, damaged, or destroyed thirty-two vehicles, thirty-two transmission towers, and thirty-eight buildings. An avalanche off Bold Peak in Chugach State Park plummeted more than six thousand vertical feet, clear-cutting 120 acres of mature timber on the valley floor. "It just literally took the forest out," park employee Jerry Lewanski told the *Anchorage Daily News*. "Nothing was left standing." This was only one of many avalanches that ran farther than had ever been observed. "We don't always know what the limits are," says Doug. "We think we do, but we don't."

CHAPTER 9

Heat of Friction

Probing during an intensive search for six buried snowmachiners that involved hundreds of volunteers over four days, Turnagain Pass, 1999 © JIM LAVRAKAS, *ANCHORAGE DAILY NEWS*

*I think that any bullet, no matter from whom it comes, is a shot
fired first to the heart of a mother.*
　　—Vietcong veteran speaking on *60 Minutes,* circa 1990

Whenever you fall, pick up something.
　　—Oswald Theodore Avery, U.S. physician

PATSY COYNE'S VOICE BURBLES WITH THE WARMTH OF A
Southern summer, far softer than the landscape she has
made her home. From her voice rise images of hot biscuits
and thick gravy, and around it wafts the aroma of sizzling
bacon. She draws out words to their utmost length, and as if
determined to impart maximum hospitality, tacks the
phrase "and that" to the tail ends of many of her sentences.
Hers is not the voice of winter except when it cracks like ice
and splinters into shards.

Growing up in Louisiana, Patsy thought she might be a
nun—until she met a boy named Bobby at twelve, married
him at fourteen, and had their first child at fifteen. Her formal education ended with the ninth grade; pregnant girls
weren't allowed in school. But Patsy already knew something
about raising babies. In the dirt yards of her childhood, there
had always been a gaggle of cousins and neighbor kids along
with rabbits and chickens, and she had minded a baby—her
youngest sibling—just months before she delivered her own
first child. Maybe, she says, she was born to have kids. By the
time she was seventeen, she and Bobby had three boys. In

nine years they had a total of six children—five rough-and-tumble boys with gentle Irish names like Keith, Wes, and O'Neal, and a lone girl named Angie.

For a time Patsy and her husband, Bobby, who was four years her senior, went to live with his parents in Mississippi. Bobby found work as a roustabout in the oil fields and in 1969 followed his brother-in-law north to work on drill rigs in Cook Inlet. Before long the oil boom at Prudhoe Bay would make Alaska the new land of promise, and eventually everyone in the family except for Patsy would do a stint on the North Slope. But in May 1969, when Patsy packed up her kids, she thought they were journeying north for the summer. She didn't realize, she says with a laugh as light as champagne, how short Alaskan summers were. Hard-tethered to the South, she had seen snow all of two or three times in her life. Moving away from the cocoon of family just wasn't done; most of her relatives had died within sight of where they were born.

But winter came in August, and Bobby saw opportunities in Alaska that were only hopes back home, so the time for migrating south slipped away. The family upgraded from their first breezy shack to another, barely more than a lean-to, with no plumbing and only the heat of a wood barrel stove. They rented the cabin for almost nothing from Curt Falldorf, who had bought a 160-acre homestead and wanted to have the place occupied as a defense against vandalism by gun-toting kids. (Twenty-one years later Doug and I would direct an avalanche search for Falldorf. His snowmachine was found first, under twenty-eight feet of avalanche debris so dense and deep that probers had rammed holes through the aluminum running boards and gas tank without noticing. It took a week to find and excavate Falldorf's body from thirty-four feet of snow.)

The first two years were "very, very hard," says Patsy, and after a visit down south, she "cried all the way back to Alaska." Because of his shift schedule, Bobby was away for weeks at a time, but when he was home, he ruled with an iron fist. From an early age, the Coyne children were disciplined to say "Yes, ma'am" and "No, sir," even to their parents. They would have caught holy hell if they'd called a teacher by a first name. Patsy grew up with her kids, playing ball, arm wrestling, tobogganing, and inciting not merely pillow fights but wars with whole mattresses. "People think the boys got their wildness from Dad," says Angie, "but it was Momma who drag raced the Bronco until the floorboards smoked."

The family moved into a trailer in Sutton, an old mining community of dirt driveways along the Glenn Highway, the winding two lanes that, at the time, formed the only overland connection between Anchorage and the rest of the world. The scenery along this stretch of road is the stuff of poetry, with the Talkeetna Mountains to the north, the glaciated horns of the Chugach Mountains looming to the south, and the braided, fast-flowing Matanuska River between the two. Palmer, thirteen miles to the west, is the closest real town, with schools and retail options other than a single convenience store. "Mom would just open the door after breakfast," remembers Angie, "and sweep us out from underfoot. We were poor. We didn't have many toys. We didn't have television. So we ran wild in the woods." Bobby prohibited bicycles because the only place to ride was on the highway, so once the Coynes had a few dollars to spare, the kids began tearing through the back hills on a ragtag collection of minibikes and go-carts. Their first snowmobile, a slow dog of a machine that all six kids piled onto at once, provided entertainment for close to a decade.

Patsy's tribe played and fought with equal passion. But no matter how fiercely they fought among themselves, they always united against insults from outsiders. The boys didn't hesitate to use their fists as weapons and, accustomed to cutting firewood and putting meat on the table, handled chain saws and guns with ease before they were teenagers. *Rowdy* is the first adjective Patsy picks to describe the black-eyed, boozy adolescent years of her three middle boys, Cleve, Keith, and Wes. The Coyne boys gained a reputation as the rough backwoods bunch from Sutton. "No one ever walked over my kids," says Patsy, with a smile that crinkles the corners of her eyes.

"Keith got in plenty of trouble," she recalls, gesturing as though fanning away the trouble like smoke, "but I always called him Sweet Bubba." Loyal and sensitive, Keith was also defiant. If someone told him not to do something, he would go out of his way to do it. Like the best friend who died in his arms after a car wreck, Keith seemed unlikely to survive his teenage years. He beat the odds, though, and once Andy and Cleve left home, took over the role of big brother to the younger pack of Angie, Wes, and O'Neal.

In high school, if Wes thought someone was messing with his girlfriend or "giving him lip," he let fly with his fists without preamble. "I'd fight for no reason and I'd fight with my friends," says Wes with an offhand laugh. "My friends were older and tougher, so for three or four years, I got beat up. But I didn't back off; it's just not in my nature to cower down." Between fights he guzzled as much alcohol as he "could get his hands on." School bus rides were a daily opportunity to sit in the back of the bus and drink. His drinking got him booted out of school for a semester, but he reenrolled because, he says, the only thing he cared about other than his family was being on the wrestling team. Anxious not to jeopardize his eligibility, he mustered enough self-control to move his fights from the school grounds to the churchyard.

As a baby, Angie was Patsy's "little doll," though she grew up just as scrappy as her brothers and proved to be the star athlete in the family. "I received a lot of sympathy for having five brothers," says Angie, who speaks in a muted version of Patsy's Southern twang. "But really it was more like being the only child in a big family. I had the benefits of playmates and constant action," and, she says with a grin, "I had a lot of power." In high school her dates were initially reluctant to pick her up at home. Once they ran the gauntlet, however, they were surprised to find her brothers as kind and funny as they could be mean and tough. Angie went on to be the second person within her immediate family and sprawl of cousins to graduate from college.

By the 1990s most of Patsy and Bobby's kids had gathered spouses and children of their own and were living just down the road in Wasilla. Little more than a railroad siding when Patsy and Bobby moved north, Wasilla has flourished with the same oil dollars that have afforded the Coynes a measure of long-sought economic security. As Nan Elliot writes in the guidebook *Alaska Best Places,* the town "boomed with strip malls during the pipeline construction years of the late 1970s, and big money was liberally laced with questionable taste. The wild feel of the place is gone. But never fear. It lives on in the woods and foothills a stone's throw away."

Hatcher Pass is one such wild retreat. With high alpine meadows that blaze with wildflowers in summer and rough-hewn black rocks framing white peaks in winter, this cleft in the Talkeetna Mountains had the majesty to make Patsy Coyne fall in love with Alaska. Only a short spin from home, Hatcher Pass afforded the Coynes three decades' worth of tramping, picnicking, sledding, skiing, and snowmobiling. Until December 1999, Patsy couldn't get to Hatcher Pass often enough. Now she can't make herself go back.

❉ ON THE DAY AFTER Christmas 1999, Keith Coyne, weeks shy of his thirty-eighth birthday, called his mother and offered to shovel the roof of her trailer in Sutton. Since Keith had given up drink for God and was a family man with two teenage children, Patsy had released most of her worry. She urged him to enjoy his first day home from the Slope; the roof could wait. With only a sliver of free time that Sunday morning before a church party, Keith loaded his brand-new-for-Christmas snowmachine into the bed of his pickup and headed for Hatcher Pass.

Having just returned from Alaska's dark northern brink, Keith knew little about the recent tumultuous weather in the southern half of the state. It was a nondescript gray day in Wasilla, but as Keith's truck climbed beyond the trees into Hatcher Pass, a potbellied sky pressed down on the ridge-lines, and the air grew thick with blowing snow. The visibility was lousy, but Keith didn't have grand aspirations. All he had time for was a quick spin.

"For all the hundreds of outings we took in Hatcher Pass," says Angie, "I don't remember ever seeing an avalanche. Knowing what I do now and given all the places we've gone riding, I can't believe one of us didn't get caught sooner." In his living room across town, Wes echoes her. "I'd seen three avalanches before Keith's accident. Now I've seen thousands. I'll be driving and I'll see one ripped out to dirt four miles off the road. I see them everywhere I go."

In winter the high country of Hatcher Pass is accessible only from Palmer. The road that continues through the pass proper and snakes twenty miles downhill to Willow (a small town north of Wasilla) is unplowed and closed to automobiles for eight months of the year. During snow season it is ruled by snowmobilers. After unloading his "sled," Keith Coyne accelerated up this road, heading toward the saddle of the pass just a few minutes of fast riding away. It would be a betrayal of snowmachine culture not to note that he was riding a Yamaha Mountain Max 700, for snowmachiners are among the most brand-conscious consumers on the planet. Even in summer a rider's loyalties are typically broadcast by the brand name emblazoned on the visor of a baseball cap or across the back of a neon-colored jacket or stickered to the rear window of a 4 × 4 pickup truck.

Just below the U-shaped pass, Keith encountered a pack of roughly twenty riders, many strangers to each other, congregated on a flat bench on the uphill side of the road. Some were laying highmark tracks onto the slope of Skyscraper Peak above them. Highmarking—trying to reach the highest point possible before turning and rocketing downhill—is also known as highpointing, hill climbing, and hammerheading. Competitive and fun, it pits engine muscle, traction, and acceleration against gravity and is a popular test of skill and daring among snowmachiners. Not all snowmachiners seek the challenge of highmarking, but those who do regard it as a summons and an inalienable right.

Highmarking is inherently dangerous because it involves

approaching steep slopes from the bottom with millions of pounds of snow hanging in balance above the rider. Most avalanche accidents involving snowmachiners occur while highmarking; in the past decade, snowmachiners have replaced backcountry skiers and climbers as the leading avalanche victim group in Canada and the United States. (In Europe, riding snowmachines for pure fun is generally outlawed, and the rest of the world has not cottoned to the sport.) Essentially, highmarking is like tickling a sleeping giant under the chin and running away, which is fine as long as the giant is known to be friendly. Highmarking in snow of precarious or unknown stability is like playing Russian roulette with bullets in most of the chambers.

The typical highmarking accident occurs when a machine becomes stuck high on a slope and the rider is trying to extricate it, either alone or helped by a second sledder. With the advent of more powerful machines and better traction, even relatively unskilled riders are now able to climb terrain that would have been unreachable only a decade ago, under a much greater range of snow conditions. One of the six men I helped dig out dead from a half-mile-wide avalanche in Turnagain Pass in 1999 was riding a snowmachine for the first time in his life. If snowmachiners would adopt the habits of riding one at a time and not parking at the base of avalanche-prone slopes, the number of fatalities would likely be whittled by at least a third, if not by half.

A good highmarker doesn't sit sedately on the seat but stands or rests a knee upon the seat. Keith Coyne was a good highmarker. He idled for a moment, but he didn't take off his brand-new helmet or stop to chat. Around 11:45 A.M., with 698 ccs of new engine under him, he thumbed the throttle and gunned the sleek blue Yamaha toward the clouds.

Keith hurtled up a broad alluvial fan and into the narrow waist of a rock-walled gully before he bogged down, roughly 150 vertical feet above where he had started. Snowmachiners know as much about getting stuck as ducks do about swimming; it is par for the course. Dismounting and grabbing hold of the handlebars, Keith began tugging the machine around through a waddle of belly-button-deep new snow, trying to get it pointed downhill. Stressed by seven hundred pounds of machine and two hundred pounds of man in the literal gut of the slide path, a small avalanche broke loose below him. For about ten seconds, it seemed just a close call—no one knew that the shear failure was propagating uphill along the unseen underbelly of the slab. Some of the other riders, though, saw the second fracture break about a thousand feet above Keith. "It started out small and then looked like the whole mountain was coming down," twenty-one-year-old Mike Anderson told the *Anchorage Daily News*. Another young snowmachiner added, "Some kid yelled, 'Avalanche!' and we all went running toward our machines. He [Keith] was still stuck up there . . . I looked back and he was ducked behind his machine. I looked up again, and he was gone."

No one was certain how many of those gathered at the bottom had been unable to outrun the avalanche, but they knew that the rider on the blue Yamaha had never had a chance to flee. Several snowmachiners raced down the mountain for help. Others stayed to search. Rescue gear was sparse, so the resourceful improvised probes from markers staked along the edge of the road. Somewhere amid the chaos, a beacon signal was heard and a twenty-minute digging frenzy ensued, resulting in several gaping holes big enough to bury station wagons. Keith was not wearing a beacon; the errant signal had to have been emitted by a would-be rescuer who had forgotten to turn his own beacon from transmit to receive. Rescuers tracking rescuers is a common snafu, especially in the early adrenaline-blurred moments of a search.

Moments before the avalanche, Alaska State Park ranger Pat Murphy had been driving to work, approaching Hatcher Pass from the valley floor in Palmer. "I looked toward the mountains," he remembers, "and couldn't see them for the wall of clouds. I thought to myself, *Whoa, this is a bad day.*" It was "windy and snowing and shitty" as he turned the corner near the end of the plowed road where Keith Coyne had left his truck. When he saw that the strong south winds were causing pluming that was stripping the ridge crests bare and heavily loading lee slopes, Murphy thought, *Oh man, this is a really bad day.*

Pat Murphy suspected trouble the moment he looked up the unplowed road leading to the pass and spotted a dark motionless knot of people. A moment later a snowmachiner skidded to a stop next to Murphy's truck, tore off his helmet, and shouted that at least one rider, and maybe many more, had just been buried in an avalanche.

Two days earlier Murphy had searched in vain amid man-high dunes of drifted snow for his own snowmachine, which before the storm had been neatly parked as usual alongside a nearby lodge. Finding it had seemed a lost cause until he noticed a fist-sized corner of the plastic windshield protruding next to his toe. It had required almost three hours of arduous archaeology to reclaim his sled and, with a screwdriver, to meticulously chip masonry-like snow from every part of the engine. Now the machine was submerged again, though not as deeply. Before Murphy picked up his shovel to excavate it, he called the Alaska State Troopers.

Trooper dispatch began marshaling additional manpower and equipment, and Murphy's radio sparked alive with the staccato of action.

By the time Murphy reached the accident site, roughly forty-five minutes after the alert, he knew that "the day was as bad as they get." Sustained winds of fifty miles per hour and gusts topping seventy-five made it hard to stand up, difficult to breathe, and almost impossible to hear any of the radio traffic. The bystanders had begun some haphazard searching, though they could hardly see their own feet. The number of those buried still hadn't been pinned down, but, stumbling and pointing, each disparate group had counted heads. Other solo riders might have been caught, but the only person known for sure to be missing was the unidentified man on the blue Yamaha.

Taking charge of a distressed, ill-equipped, blizzard-battered band of volunteers who have differing accounts of what they have witnessed, and who are largely strangers not only to one another but to avalanche rescue, is like climbing into a clothes dryer while it is tumbling and attempting to organize all the socks and underwear. Pat Murphy, a ranger for more than fifteen of his forty years, was trained to try. After encouraging those who weren't wearing avalanche beacons to leave for their own safety, he directed those who elected to stay. They probed for two hours, during which Murphy began to hear his own doubts over the shriek of the blizzard and the urgency of the moment. Only a fraction of the starting zone had released. Higher on the mountain, the winds were likely pegging a hundred miles per hour and were cross loading the slope above the group with wind slab that posed a greater threat by the minute. Darkness was approaching, which would not only make it difficult to monitor the escalating hazard, but would complicate the departure of physically and emotionally spent searchers. Even over the garbled static of the radio, Murphy understood that he was getting increasing pressure from the evolving pyramid of command to bail.

Murphy and his bedraggled troops had retreated only a short distance when it occurred to him that the winds were rapidly rubbing out the accident site. If any future searching was to have a prayer of success, he needed to mark the boundaries of the debris pile with tall, firmly planted wands. His request for volunteers, he remembers, was met with a response "like a scene from *The Three Stooges.* Everyone stepped backward and looked every direction but toward me. No one wanted to go." Murphy drafted two men, and the three of them trudged back uphill, dragging wands. When they reached the lower perimeter of the avalanche, they split left, right, and center to mark the debris, but before they could accomplish the task, another slide poured down, unseen and unheard, stopping just short of the hapless helper on the left edge.

The lot where Keith Coyne parked had been crammed almost to capacity with trucks and snowmachine trailers— holiday weekends are long-standing invitations to play, weather be damned, and the day after Christmas is prime time. When rescue coordinator Paul Burke arrived in the parking lot, he encountered a woman from the local ambulance service standing beside her rig. Dressed in cotton stretch pants, a light windbreaker, and low-cut shoes, she said she was in charge. Her intention was to send five medical technicians up the road with no beacons, no probes, and a couple of shovels. "Thank you but no," Burke responded in his typical manner—part cop, part choirboy. "The wind was horrendous and avalanches were rolling down everywhere," he recalls. "That crew might as well have gone looking for victims of the *Titanic* with snorkels and flippers."

Burke's job was to assume command and get the right people moved into position; it was he who would summon me to Jerry LeMaster's avalanche in Cordova exactly four weeks later. The first imperative was to empty the parking lot to see how many trucks remained unclaimed, and thus try to confirm the number of missing. Ordered to leave, snowmachiners packed up their gear and began filing down the mountain, a wink of taillights as they cautiously descended the slippery road. While the evacuation was in progress, an avalanche pounced to the road but fortuitously covered only the uphill lane.

In the end, aside from the rescue vehicles, a lone white Ford F-150 pickup sat in the lot, shaking in the gale. Angie's husband, Ed Reavis, who had gone to play elsewhere in Hatcher Pass that day, identified it as Keith Coyne's even before the Troopers ran a check on the plates. After hearing snatches of the witnesses' accounts, Ed and some of Keith's friends provided the link between the white truck and the blue Yamaha. Ed called Angie, and Angie called Keith's wife, Heidi, who had already received calls from both the Troopers and her church prayer chain. The call to Patsy came while she was with O'Neal, watching his son play hockey. Patsy summoned Bobby and Cleve home from the oil fields.

Angie also called Paul Burke, who had by now established a rescue base in the rambling Motherlode Lodge a few miles down-mountain from Keith's truck. They were friends from church, and Angie trusted Paul, knowing him to be exuberantly optimistic and kind. When she asked for the straight scoop, Paul told her that he was 99.9 percent sure that Keith was dead. "Still," says Angie quietly, "not knowing was as bad as knowing."

Seventy miles away Doug and I started moving the instant we heard from Paul, who had called us as soon as he received word of the accident. But nothing about the next two days would go fast enough for anyone. The winds at our house hadn't dipped below sixty miles per hour for a week, while the temperature had turned so unseasonably mild that snow had given way to rain. Usually, we can find a speck of gravel or a bit of soft snow toward the edges of our road to help the studded tires find traction, but the road had mutated into an icy bobsled run made even more treacherous by a sheen of water. We couldn't hope to go anywhere until we wrestled chains onto all four wheels of the Suburban. At the last moment, I grabbed our dog; we wouldn't be making it back uphill to our house anytime soon.

We had only crested the top of the driveway when a linebacker of a gust slammed our three-quarter-ton vehicle broadside and sent it into 360-degree spins that would have put an Olympic figure skater on the medal stand. The dog whimpered from the floorboards while I closed my eyes and Doug fought to keep the truck from plunging over the edge of the hill on one side or into the ditch on the other. Two complete revolutions later, we were alive at the bottom of our hill, greeting life with the enthusiasm of the condemned granted reprieve.

We arrived at the Motherlode about the same time as a wind-blasted Pat Murphy came down off the mountain. The lodge had been gearing up for a wedding, but the owners graciously offered the largest dining room as a war room for the rescue operation. We considered the immediate options. Sometimes, we knew, law enforcement officials are too quick to decide that darkness precludes continuing a rescue effort. Searching at night is inconvenient but possible. Every elapsed minute reduces a buried victim's chance of survival, and there are a lot of minutes in a night. Even horrendous avalanche conditions may not be reason enough to delay. Once, in Colorado, Doug and I stood with members of the Aspen Mountain Rescue Group being stonewalled by a representative of the county sheriff, who kept insisting that conditions were too dangerous to let us search for the occupant of a tepee that had just been demolished by a large avalanche. We had spent the previous three days training members of the group in the same area and had grown intimate with the conditions. The snow was so sensitive that we'd been able to make suspect slopes avalanche from more than a quarter mile away simply by jumping on flat surfaces. In a strange role reversal—because we are usually the ones championing the safety of the rescuers—Doug and I argued that given the extreme instability of the snow, everything that could possibly avalanche might very well have already done so. But several days elapsed before the sheriff allowed anyone past the blockade to observe that, indeed, every path that posed a threat had long since avalanched. Found pinned against his woodstove, the victim had lived for a few hours.

In Hatcher Pass, with three to six feet of new snow and winds reloading the slopes almost as fast as they could avalanche, searching for Keith Sunday night was out of the question, although his anguished brothers considered bypassing the Trooper roadblock by leading a posse up and over the pass from the Willow side. The more significant question for Doug and me was whether we could make the slopes safe enough Monday morning to allow access to the accident site.

Patsy Coyne was one of the few among the family quickly gathering in Keith's house who didn't just hope but truly believed Keith was still alive. "I knew his strength and determination," says Patsy. "Right up until the end, I thought he'd be there, mad as hell that it took so long to find him." Her family was reluctant to let her leave their clutch Sunday night, but Patsy insisted upon going home. Unable to take refuge in sleep, she began to cook. She peeled potatoes and soaked them in cold water. She made spaghetti and a vat of meat sauce to fuel dozens of wet searchers the next day. By the time she turned her attention back to the potatoes, deftly dicing them into soup, Keith Coyne had been buried more than twelve hours.

Unbeknownst to Paul Burke or us, the family had been advised by a trooper that the rescue would recommence at 9:00 on Monday morning, a remarkably optimistic and misleading prediction. In the meager December daylight, it wouldn't even be possible to see well enough to do the necessary avalanche control until almost 10:00 A.M. After pumping the eyewitnesses for information late Sunday afternoon, Doug and I had made chains of phone calls arranging the logistics for two possible contingencies. Plan A was to load a helicopter with explosives and have it positioned at the base of Hatcher Pass by first light. This would allow us to assess the hazard from the air and, if need be, do a strafing run in one fast, efficient swoop. Plan B involved a more logistically cumbersome artillery attack. This would be brought into play only if the winds were still too strong and/or the light in the starting zones was too flat to allow safe flying. Bombing in gusty winds is a one-way ticket to trouble; a colleague once lit a charge and threw it out the open door of a helicopter, only to have a blast of wind sling the bomb back in. A mad scramble ensued to find the sizzling charge, which had slithered under the backseat amid cartons of highly concentrated explosives. The ninety-second fuse that would detonate the blasting cap had less than twenty seconds to burn when the bomb was kicked back out the door.

It was midnight before we went to bed Sunday, and I was up to check the weather on the computer at 4:00 A.M. With snow falling and the winds still screeching, Plan A was a pipe dream. We made the calls necessary to put Plan B into effect and geared up for the day. For all our planning, though, we were completely unprepared for what lay in store.

※ THE COYNES arrived at the Motherlode well before 9:00 A.M., staking out the restaurant and bar as a waiting area and receiving station for a volunteer army of friends. "The bar was open early and people were taking full advantage of it," recalls Burke. Sensing a primal vigilante mood, he asked Angie and her husband to serve as intermediaries between the war room upstairs and her parents and their supporters downstairs.

Outside in the dark, Wes, Cleve, and O'Neal sat on their sleds anticipating immediate action. But by 9:00 A.M. all that appeared to be happening was that a trooper stationed at the roadblock was preventing rescuers from getting to Keith. "I had a picture of how it should go," says Wes, "and it seemed like we just got handed one lame excuse after another." First, he recalls, "we were told we were waiting for the cannon." The idea that more avalanches might be dumped on top of Keith killed the last embers of hope that he might be found alive. Wes says that he can't begin to describe his state of mind that morning, but when he tries, it sounds like anarchy. "I didn't care about making the area safe. I didn't care if anyone got hurt or even died. I didn't care if anyone helped us. I was ready to do whatever it took to get my brother off that mountain and back to Momma."

When towing a nearly 5,000-pound 105-mm howitzer a hundred miles over icy roads, it is not prudent to hurry. Three Department of Transportation avalanche workers were at their Girdwood base hours before daylight to check that conditions would permit them to leave the Seward Highway, which was their primary area of responsibility, defenseless for the day. They prepped the gun and loaded the ammunition, which by regulation must travel separately from the weapon, into a second truck. As they drove north, they found the margins of the highways littered with vehicles, but the highways were easy street compared to the winding seventeen-mile access road to Hatcher Pass, which was coated with ice. Worried that the trailer would jackknife, the crew stopped to chain up, groping their way around the truck to keep from falling. Ahead of them, another transportation department crew had risen early to fire up heavy equipment and wrestle the steep, drifted hairpin section of road above the Motherlode into something manageable enough for the howitzer and rescue vehicles.

By the time the gun arrived at the Motherlode around ten, the sky had lightened to a sludgy gray, revealing fat sausage-shaped wind clouds scudding along the ridges. Wes and his brothers, still hunched on their snowmachines, were seething, their eyes like drill bits and their faces glowering beneath the lifted visors on their helmets. Their speech had become a nearly unbroken string of invectives, and joining the usual list of expletives were my name and Doug's. They had been told that avalanche control could not take place until the starting zones could be seen. While they were forced to bide time, the skies were lowering and thickening with more snow, which only served to heighten their anxiety.

Doug and I went through the roadblock and back uphill toward Keith Coyne's truck with the gun crew. We were asking them to shoot the slopes hanging above both the unplowed road to the pass and the accident site itself, backcountry areas that had never been fired upon by artillery. The trick was to find a level-enough place on the plowed road to set up the gun without getting so close to the flanks of the mountain that the shooting angle would be too oblique. Range was fortunately not an issue, as a howitzer can easily find its target from four miles away.

The crew couldn't shoot until we were sure that no one had wandered from the Willow side of the pass into the firing zone. Pat Murphy and another ranger had volunteered for the edgy duty of running sweep up to the pass on their snow-machines. The trick to maximizing safety is to move as fast as possible, but Murphy and his partner hadn't even reached the accident site before they were thwarted by the debris from a fresh avalanche that had buried three hundred feet of the route. "The debris was six to eight feet deep, and the blocks all had sharp, crisp corners," says Murphy. "Given that the weather was filling our tracks in almost as fast as we were making them, the avalanche must have come down within the hour, two at the most. If we'd had eighty people moving in to search, we would have had a disaster on our hands."

In light so flat that up was indistinguishable from down, Murphy broke a trail over the avalanche rubble, making it past Keith's avalanche and up to the pass without encountering any stray travelers. Originally, he'd planned to drop over the other side and wait out the shooting in a completely safe area, but the slope he needed to descend was bulging with so much wind slab that he figured he'd turn himself into an avalanche victim if he tried. Instead, Murphy and his partner turned their machines broadside to the wind funneling ferociously through the notch and took cover behind them. Keying his mike, Murphy repeated several times that he was clear and firing could commence. If there was a response, he couldn't hear it. What he didn't explain, because communication was too difficult and he was doing his utmost to keep from getting blown off the mountain, was that he wasn't quite as far away from the slopes as he was expected to be.

The crew sighted the nine-foot barrel of the gun into the starting zone of the path closest to the juncture of the plowed and unplowed roads and loaded a four-foot-long "bullet" packed with 4.8 pounds of TNT. "Clear to the front!" yelled the assistant to the gunner who would pull the trigger. "Clear to the rear! All clear!" (Unlike recoilless rifles, which have not been widely used for avalanche control since defective rounds exploded in their barrels in 1995 and 2002, howitzers don't emit a tongue of flame into a 100-foot "kill zone" behind them, but extra precaution is always taken.) Knowing what was next, everyone in our small group clapped hands over headsetted ears and took a breath, bracing for a hard thump against our chests. Doug and I had left the windows of our nearby truck slightly open so that they wouldn't shatter in the concussive reverberation, but there hadn't been any way to warn our still-captive dog that he was about to be rudely awakened from his nap. "Ready to fire. Fire!"

The bullet began its long, unseen arc toward the mountain. When the round hit, the slope blackened, though the thud of impact was lost in the roar of the wind. No avalanche. Swinging the barrel of the howitzer west up the road toward the pass, the crew sighted in another target. No avalanche. Another shot. Again, no slide that we could see, though the murky light left us squinting with doubt.

In addition to the corps of friends the Coynes had assembled, the members of no fewer than six volunteer rescue groups had taken time from paid jobs to be on hand at the Motherlode on Monday morning. Each person milling about was an affront to the Coyne brothers' sense of urgency. Monitoring the shooting over the radios clipped to the hips of waiting searchers, they thought the point had been proven. It was past eleven; how much more time were we going to waste? Six shots into the routine, I was also beginning to wonder, though the shots were drawing closer to the slopes near the pass that had borne the brunt of the wind loading and were likely to be ripest. I'd heard enough by now to know that if we didn't get any releases and, suspecting instability, still decided to limit or postpone the search, the Coynes' simmering hostility would escalate to full-scale war. I was prepared to argue on their behalf that they and any willing friends should be loaned the necessary equipment and allowed onto the mountain; it was their prerogative to risk dying. However, they needed to understand—and make starkly clear to Patsy and others—that should they run into trouble, no rescue would be launched for them if the danger persisted.

Avalanche control breeds tenacity, even mulishness. Once when Doug and I were helicopter bombing to secure an area for a power-line repair crew, we threw hundreds of pounds of charges, fifteen to thirty pounds at a time, without result. As we contoured up the mountain, railroad and highway officials pressured us to quit so that they could get on with the business of opening corridors that had been sealed for too many days. Doug, more impervious than me to second-guessing, insisted that we stick with the plan. When we reached the top of the mountain, we had just one five-pound charge remaining. It found the "sweet spot," ripping out a slab nine to twelve feet deep and over a half mile long. The avalanche generated churning powder clouds that flew down four separate chutes and ripped out acres of mature timber. Had the avalanche released on its own volition, it would likely have buried the repair crew. The new slide deposited another twenty-five feet of debris on top of the fifty-foot pile from the avalanche that had damaged the power line.

In the pass Pat Murphy didn't know where or when the shots were going to hit, so he watched for the telltale puffs of black smoke. He was close enough to hear the boom of the bullets as they hit the mountain and ruptured into shrapnel, particularly when two shots hit a rock protuberance known as Nixon's Nose that was disconcertingly close to his perch. It was the last two shots of twelve, however, that triggered such a rush of adrenaline in him that he almost felt warm again. Both shook loose avalanches in the path where Keith Coyne was buried. The second of these was about the same size as the avalanche that had caught Coyne but began higher and to the left. Given the intense rate of loading in the bowl, it is virtually certain that this avalanche would have released naturally during the day. From the gun we could only see the slab dislodging as though parachuting slowly from the clouds, but Murphy saw it rally speed, pour through the gully where Keith had been stuck, and spill across the accident site, adding more debris to the original pile and razing many of the markers.

From my point of view and Doug's, another needless tragedy had just been averted and we had bought some hours to search without fear of ambush. From the family's perspective, however, we had just put a nail in Keith's coffin.

❄ OF THE EIGHTY searchers amassed and dressed in storm gear, a few would stay at rescue base to execute logistics worthy of a military campaign, and some would operate a motor pool, ferrying people, equipment, and food to the site on snowmachines or in snowcats. Most would spend the day systematically pushing metal probe poles through the snow and advancing uphill a step at a time. In the crowd Paul Burke had spotted tall, sturdy, levelheaded Kevin Siegrist, a volunteer member of the National Ski Patrol. He pulled Kevin aside, saying, "Boy, have I got a job for you."

He assigned Kevin to be leader of Team 1, made up entirely of the Coyne brothers and their friends. They would go in first, along with Doug and me as accident site commanders, a few of the witnesses, and two search dogs and their handlers. During the delays of the morning, Kevin strained to keep them from storming the roadblock and tried to make good use of time by teaching the group how to use avalanche beacons and run a probe line. But probing a parking lot while Keith lay buried on the mountain drove them wild with frustration. How many more excuses and rules were going to be shoved in their way? Kevin didn't want to force the point that Keith had been buried for twenty-four hours and had only a shade more than zero probability of being alive. He patiently fielded a barrage of increasingly critical questions, even though, he says, "they didn't understand and they didn't want to understand." At the end of the day, he would tell Paul, only half joking, "Next time you issue me a radio and send me up a mountain with a team like that, you can issue me a gun too."

Just before noon we were at last moving toward the accident site along a route that Murphy had flagged with wands. Originally, he had set the wands about forty feet apart, but as the snowfall intensified, he narrowed the spacing to less than ten feet, and even that was sometimes barely adequate to keep us on track. I rode to the accident site with Wes Coyne.

When it comes to riding on the back of a snowmachine in whiteout over bumpy terrain, I am not at all scrupulous. I wrap my arms around the waist of whoever is driving, even if I've never met him before and he is beginning to hate everything about me. I cinch in tight, clamp my legs around his thighs, and press my chest firmly against his back. In my experience, if we ride and lean as one, we are more likely to arrive as two. Rescuers who disdain such intimate body contact inevitably end up getting thrown off at least once. I clung to Wes for all I was worth, and that would be the last strong solidarity with the Coynes I'd feel for quite some months.

In a search, figuring out where the victim is most likely to be buried and thus which areas should be assigned priority is as important as having the proper tools and technique. When I first arrive on scene, I try to ignore the constant squawk of the radio and the hubbub of the incoming teams for a minute or two and just look. Often I'll climb a short way up the debris to gain some physical space and perspective. It feels as though I am peeling my mind open and, with a slowly revolving third eye, taking as wide a view as I can. A common rescue acronym is STOP: stop, think, observe, plan. What do I see? Are there entry tracks that can help determine the trajectory of the victim? Are there any surface clues, like the ski of a snowmachine or the heel of a boot? What are the flow characteristics of the avalanche itself— that is, was the victim likely to be swept left, right, or straight downslope? Where are the likely catchment areas, such as benches or concave traps? Is the victim's dog drawn to one location? (If he is a loyal, smart, well-loved dog, he may be sitting on top of his master.) What assumptions am I making about where to search? How can I most efficiently allocate the available resources? What other resources do I need or am I going to need three, four, or five hours down the line? Every search has different answers. During rescues, even amid constant clamor, my thoughts seem to arrange themselves quietly, like sequential frames of a slow-motion movie. The fragments of information begin sifting as if each has tangible weight, until they slot into patterns. If I could think so calmly and logically during ordinary life, I would be very grateful.

Failing to take witnesses back to the accident site is like trying to play a game of chess blindfolded. A witness may know things he doesn't even realize he knows. Where was he at the time of the avalanche? Was he caught? If so, where did he end up? What was his position relative to the missing victim at the time the avalanche broke? Where was the victim when last seen? Was he still on his snowmachine (or skis or snowboard)? What direction was the victim moving? Does the helmet found on the surface belong to the witness or to the missing person? How about the glove? It can be hard to get straight information from someone who is rattled, cold, or exhausted. Compounding the confusion, the avalanche has usually completely transformed the landscape: deep gullies may be filled in with snow; hills exist where there were none before.

If Doug and I are working together, then often while I am taking my wide view, he will take a much narrower one and work separately with each of the witnesses, getting them to slow down their narratives so that he can draw out the details. If possible and practical, he will have witnesses return to the spot where they were standing at the time of the avalanche. Often Doug will encourage them to close their eyes and describe what they saw, and then have them open their eyes and find the bearings they have just detailed. In this way, Doug once helped a son to realize that the "big rock" he had hurriedly pointed to as the last place he had seen his father was the wrong rock—the actual landmark was two hundred feet away, and searchers were probing where his father could not be. If there is more than one witness, their stories must be carefully held to the light and compared. When there are multiple witnesses and multiple victims, pinpointing the crucial information can feel like being lost in a very large city while bystanders shout contradictory directions in sixteen languages. At Turnagain Pass the spring before Keith Coyne's avalanche, I'd grabbed the box holding a twelve-pack of soda off the back of a snowmachine, flattened the wet cardboard, and diagrammed each piece of intelligence in order to untangle the web of details enough so that we could mount the search for six men buried in an area equivalent to about twenty-five square city blocks.

If you climbed to the area where a victim was last seen and dropped a hundred basketballs, some would roll relatively straight down the fall line, while others would deflect left or right. Together, their trajectories would describe a rough triangle, a "cone of probability" that delineates the primary search area. If a victim was caught high on the slope, near the fracture line, he or she is likely to be buried in the upper third of the debris, although all bets are off if the victim stayed on his skis or machine for a while, trying to outride the snow. Statistically, most victims are found in the toe of the debris. Sometimes there is a trail of clues that makes it possible to choose the most likely search area with confidence. When there is less to go on, the search can feel like looking for a kernel of corn buried in the desert. The effort to find Keith Coyne fell into this second category.

All we knew with conviction from witnesses is that when they had last seen him, Coyne had been in the center and near the lower end of the gully, standing to the left of his machine, hanging on to the handlebars. There were no surface clues and no beacon signal. With many of Murphy's wands buried by the avalanche we had just shot down, we no longer knew the exact boundaries of the accident site. Even the original fracture line had filled in overnight. What lay in front of us was a jumble of hard angular blocks strewn across an area roughly 150 feet long by 150 feet wide. When I pushed my probe through the debris, I could feel a slight difference in density between the new debris and the old. My estimate was that there was roughly four to eight feet of new debris on top of the first avalanche, which seemed quite a bit deeper. Even when I screwed extra sections onto my probe and pushed it down twenty feet, I didn't penetrate through the avalanche rubble into the softer original snow surface. My hunch was that Keith was midway or lower in the lobe of debris, and from the lean of the terrain as I looked upslope, I suspected that he had been carried to the right of center. Indeed, this was where he lay hidden, though it would be hours before we knew it.

Most searches that go on for any length of time generate entreaties from family and friends of the victim to bring in high-tech tools like ground-penetrating radar and infrared sensors. Generally, such equipment accomplishes little other than to make people feel as though everything possible is being tried. I've even seen state-of-the-art military magnetometers that can supposedly detect a garbage can at sea level from an elevation of ten thousand feet pass five feet over buried snowmachines without so much as a blip. Sometimes the special requests swing more toward the spiritual. After an unsuccessful weeklong search in southeast Alaska for a woman lost at sea had been suspended, a psychic had a vision in which the victim was very weak but still alive and sitting against a Sitka spruce tree on the opposite side of the large island from where the searching had been concentrated. Logic said this couldn't be so, given the prevailing currents and winds in relation to where the boat had gone down and where the body of the first victim (her boyfriend) had been found. As nothing more than a courtesy, rescuers were sent to the far side of the island, where, ten days after the accident, they found the woman in the location and condition predicted. What we don't know about the workings of our world still outweighs our knowledge; it seems wise to stay receptive to all possible tools. Sometimes that principle puts my skepticism to the test, as when a psychic friend of one snowboarder's family asked for the name of the search dog working on-site so that she could put her dog in telepathic communication. Still, among a dazzling array of resources, from the technical to the mystical, two simple time-tested tools offer the best chance of finding a completely buried avalanche victim who is not wearing a beacon: an avalanche probe and a well-trained search dog.

Sending the two dogs and their handlers to the Coyne avalanche with the first wave of searchers would give them a chance to work before the debris got too busy or contaminated, though search dogs are trained to ignore the commotion and to concentrate on scents wafting up from beneath the snow. With the debris presenting me with an essentially blank slate, my hope surged when I saw Paul Brusseau and his Australian shepherd, Chili. Now in his forties, Brusseau is a slender, thoughtful man with piercingly clear blue eyes that seem a conduit from his heart. Chili has remarkably similar eyes and is even more handsome, her silky reddish coat marbled with white and beige. At home when she hears Paul's pager go off, she gets up and waits by the door.

I've learned to place more faith in some dog-handler combinations than others; the limitations generally lie less with the dog than with the handler's ability to keep up with the dog and interpret important signals. Maybe because Paul was number five in a lineup of nine siblings, with the girls outnumbering the boys, he learned how to listen and observe, a talent he has put to extraordinary effect with Chili. "Dogs are usually not liars," says Paul, "and their body language speaks volumes." Every nuance of the way Chili scratches and wanders and sniffs means something to Paul, and not infrequently he gets down on all fours just to see the world from her perspective. One thing he will never be able to do, however, is to smell as well as Chili does. The average dog has some 220 million scent receptors in its nose, while humans have a mere 5 million. If the membranes lining the inside of a dog's nose were laid flat, they would cover several king-size beds. As British veterinarian Bruce Fogle observes in *The Dog's Mind*, "Odors play tunes, as it were, in the dog's nose." Chili has found adults lost in the woods and plane crash victims scattered across a mountainside and concealed by new snow.

Paul and Chili don't swagger into a rescue site but slip unobtrusively to work. Paul has trained Chili to think for herself, and, as usual, at the Coyne avalanche he followed at a distance, watching carefully as Chili clambered nimbly over the jungle gym of debris blocks. But as Paul says, "It just wasn't a dog-friendly day." Chili never doubled back to prod her nose deeper into the snow, she didn't stab the snow surface with her front paws, she didn't bark, she didn't wave her tail like a flag, and she didn't ask for the ball that is her reward. "Sometimes the magic is there, and sometimes it's not," says Paul. At Hatcher Pass it was still so stormy that the winds must have been wicking away any tendril of scent as soon as it reached the surface, and the air, especially at Chili's level, was a fusillade of blowing snow. Neither Chili nor the second search dog ever gave alerts to indicate that they'd found areas of special scent. Within a few hours, they were too wet and weary to continue, and their handlers had to pull them out of the hunt.

The German word for avalanche probe is *Lawinensonde,* which means "avalanche-sounding rod." A crude method, probing is how people looked for avalanche victims in the Alps hundreds of years ago. During the initial stages of a search, an individual may spot-probe here or there to check out likely areas around clues or obstacles, but probing is most efficiently done in horizontal lines using a grid spacing that balances the probability of finding someone against speed. On average, it takes twenty searchers four hours to coarse-probe an area roughly twice the size of a football field, using a spacing of thirty inches between probes and twenty-eight-inch advances. This yields roughly a 70 percent probability of finding the victim on the first pass. The probability could be increased to nearly 100 percent if the spacing were reduced to twelve by twelve inches, but probing the same area would then consume sixteen to twenty hours. Combing this fine is rarely done; instead, it is more efficient to run multiple coarse-probe lines. Just because an area has been probed doesn't mean a victim isn't there. At Turnagain Pass, where up to four hundred people probed for four straight days, several of the victims were found after making four, five, even eight passes over the same area. Sometimes the probes missed the victims, but more often inexperienced searchers numb to the tedium of the task didn't realize that they had struck a body. Depending upon the hardness of the snow blocks, probing can take a fair bit of muscle. The desired motion is less spearfishing than the application of steady force. As Doug says, "The golden rule of probing is 'Probe unto others as you would have them probe unto you.'" While many fret about injuring the victim, this is usually a false worry, given that nearly 90 percent of victims found by probe are recovered dead. The drawback of probing is simply that it takes so long.

Probes down, probes up, advance. Probes down, probes up, advance. Seventy people now, in teams of ten, hooded heads bowed against the wind and snow, the commands spoken by each probe line leader repeated like a mantra. Wes and his team chafed, stallions put to work as mules. They wanted to find Keith, and there had to be a better way than looking for him with the equivalent of toothpicks. They fired question after question. Weren't there any dogs around that could find something more than their way out of a paper bag? Why were we probing here instead of there? Why use probe lines at all? Swear words did triple duty as nouns, adjectives, and verbs. Doug and I and a growing host of others no longer had names; we were simply "fucking assholes." In the brothers' view, we couldn't be trusted and were nothing more than an obstacle to Keith's rescue. If we had known what we were doing, the sixth victim at Turnagain Pass wouldn't have had to melt out in the spring.

There were three snowcats at work, which is at least two more than normal. Wes and company knew a lot about heavy equipment; many of them were professional operators. They wanted to pull the operator out of one snowcat, which was being used to thin the deepest snow in the primary search area within range of the probes, and drive it themselves. Thirty years' experience or not, clearly the guy didn't know how to drive. They wanted a D8 Cat up there. They had access to lots of machinery. They could dig up the whole mountain if they needed to. When a snowcat broke down, it seemed to spew gasoline on the fires of their fury.

Wes found Doug on the debris, put his face within inches of Doug's, and with clenched fists began screaming about the need for more heavy equipment. Doug didn't disagree; he'd put in the same request to the logistics team the night before. But Wes was like a volcano, unable to stop venting, his rage and sorrow erupting in clouds of steam. He had hit his limit of trying to bear the unbearable. Kevin Siegrist stepped aside for privacy, changed radio frequencies, and told rescue base that a law enforcement officer was needed on scene. For the first time in all my years of avalanche rescue, I'd already done the same. Earlier in the day, a trooper had warned Paul Burke that he thought the building belligerence was likely to make arrests necessary. "Do what you have to do," Burke had replied. "I'm right there with you."

Except just now Burke was ensconced in the warm lodge, and Wes was emphatic in his intention to ride back down there, have words with him, and round up a cavalry of bulldozers that would do the job. Doug, who thought it beneficial both to have more equipment and to get rid of Wes, was more than happy to let him go. But before Wes could get his snowmachine started, he was approached by Soren Orley, a longtime member of the Alaska Mountain Rescue Group who normally lives placidly as a certified public accountant. Soren, a large blond man of stolid Scandinavian descent, is known for his equability and good cheer. Soren was hoping to make peace and to ensure that Wes rode with a partner, but in his approach Wes read a declaration of war. Wes threw down his gloves and helmet, shoved Soren in the chest, and threatened to punch him flat if he didn't back off. Blows were prevented only by the intervention of a third man, who inserted his body between theirs.

Patsy had known her sons would be "rowdy." She'd chosen to hold vigil at the Motherlode rather than at home partly because she thought she might have to ride up the mountain to calm them, though she is a big woman and hadn't been on a snowmachine in years. Asked how she could accomplish what no one else could, she didn't hesitate in her answer: "When Momma says something, the boys know it has to be." Angie agrees: "Momma is still the boss of us." Ultimately, it was the fear of doing anything to hurt Patsy rather than the threat of arrest that helped keep Wes and his brothers out of jail, though even that might have fallen short if law enforcement officials hadn't afforded them generous slack.

Wes and company had made it clear that they had no intention of leaving the mountain without Keith. But most of the volunteer searchers were Good Samaritans who had stood within range of verbal abuse all day, and more than a few had been threatened. By dark everyone was wet to the bone, and the snowfall was intensifying. With the help of generator-powered lights, we were still probing, but the energy reserves of the troops were just about on empty and, given the low morale and high level of disgust, I knew that few would choose to return the next day. I'd been fielding radio traffic all day when around 5:30 P.M., I finally heard a request I welcomed: "Jill, can you come check this out?" A prober on a team working the deep side wall of one of the trenches cut by the snowcat had a possible strike. He left his probe in place, and as soon as I slid mine alongside and felt it come to a subtly cushioned stop, I knew we had found Keith Coyne.

Keith had originally been buried under eight feet of snow, and the second slide had added roughly another five feet. To reach his body, the group quarried a hole fourteen feet deep and almost forty feet wide. Keith was on his back, with his head downhill and both arms stretched toward the sky. Given his deep burial and the absence of even a small air pocket or ice mask in front of his face, it is certain that he didn't live long enough to take a breath. His helmet had taken a hard hit from the tumbling snowmachine, so it is even possible that he died of a broken neck, but an autopsy was never done. Like protective grizzlies, the Coyne brothers and their friends growled away outsiders, safeguarding Keith's body. The rest of the group offered no resistance. We were so ready for escape that there was a virtual stampede down the hill.

Normally, the term *heat of friction* is used to describe the energy generated by the collision of snow particles during an avalanche. It is the dissipation of this heat when the avalanche comes to rest that causes the particles to fuse almost instantly, making the deposits so hard that even shallowly buried victims can rarely free themselves. Heat of friction took on a whole new meaning for me during the Coyne search. I left Hatcher Pass so well armored by righteous anger that I felt completely buffered from the usual sadness.

Often after a rescue mission, Doug and I end up fighting about something trivial, out of frustration at watching repeats of virtually the same accidents, with the same foreseeable tragic results. But weary of conflict, we came home from Hatcher Pass and made the kind of love that needs no words. Keith Coyne was the first victim in years who didn't keep either of us awake at night. For the first time ever, I felt liberating distance from a family's grief. If I saw a member of the Coyne family again in fifty years, I thought, that would be too soon for me.

❄ THE USUAL SWIRL of opinions found their way into the newspaper after the accident. One witness to the avalanche wrote indignantly to the *Anchorage Daily News,* "Mr. Coyne did not trigger the slide. His machine was silent as he was digging himself out." Even proximity had not debunked the cinematic myth of the noise-triggered slide—per usual, the weight of man and sled had tipped the balance and set the slope in motion. I tend to see ignorance as a summons, but this time it was merely cause for a good laugh. I heard that the Coynes and their supporters had practically made voodoo dolls of me and Doug and were taking every opportunity to denigrate us, but I felt no need to mount a defense. Later in the winter, Cleve Coyne would tell the paper, "As soon as Keith was found, he was a statistic. Things were folded up and gone." He was right. Keith Coyne had no place on the shelf of my heart.

The desire to understand the why and how of what happened is a common reaction among surviving friends and families, though normally I encourage them to allow the salve of a little time before they seek to learn more about avalanches. When I discovered that Wes and Cleve, along with Angie and her husband, had registered for a one-day avalanche awareness workshop to be taught by Doug and me two weeks after Keith's death, it wasn't concern for their tender feelings that disturbed me. And when I saw that a record 350 people would be attending, many of them friends of the Coynes, I wondered whether we would have a lynch mob on our hands.

I almost didn't recognize Wes when he walked into the room. He seemed so much smaller than when he'd been spitting curses into my face. Angie had been at the Motherlode while I'd been on the mountain, so we'd never met, but she looked kind and intelligent, and my stomach did little flips throughout the day when I'd catch her blinking back tears. Sure that the brothers and their friends couldn't sit calmly through eight hours of training, I braced myself for outbursts that never came. Patsy wasn't ready to come that day, but another mother of a victim was in attendance—Aedene Arthur, whose son had been killed in Turnagain Pass the previous spring. "The recent loss of Keith Coyne . . . brought it all back," she told an *Anchorage Daily News* reporter.

During the brief media blitz following the accident, the Coynes had hated us all the more for making it clear that Keith Coyne was a participant in his own death. "He didn't know any better," they'd argue, which was true. Over time, though, led by Patsy's determination to "make something of Keith's death and spare other families such awful pain," the Coynes have taken on the harder truths as a cause. Aedene Arthur and the Coyne family became the catalysts behind the formation of a volunteer group dedicated to providing emotional support to survivors, promoting avalanche awareness, and facilitating rescue response. Explaining his motivations in an *Anchorage Daily News* article headlined BORN IN GRIEF, Cleve Coyne noted that while many snowmachiners had vowed to get training and equip themselves with beacons, shovels, and probes immediately after his brother's death, since then "they haven't given it a second thought."

The Coynes have become evangelists for avalanche education, though Patsy can't bear to watch footage of avalanches in action. They have made their apologies to me and to others. Although I would have thought I could make myself rich betting against it, Wes now participates in avalanche rescues as a member of the team.

In an article titled "What Is It Like to Be a Bat?" Thomas Nagel explores the philosophical problems of imagining what it is like to be something or someone that you are not. Four years after Keith's death, Nagel's discussion came to mind both as I listened to Wes describe the agony of the search and again on the cloudy afternoon when I met Patsy for tea. She said that she sometimes sees Keith in the movements and gestures of others and even in the grandfather clock in Angie's home that chimes like a bird. As she spoke, tears came and went like rain showers, and in the burst of sunshine that is her smile, she joked that her "tear ducts are too close and need to be sewed up." "No mother should outlive her kids," Patsy quavered, and as she began to cry, I noticed that my own cheeks were wet.

CHAPTER 10

Faces in the Dark

Hard slab avalanche that caught six snowmachiners, East Fork of the Matanuska River © DOUG FESLER

Once upon a time she had known more and wanted to.
—Toni Morrison, *Beloved*

There's a period of life when we swallow a knowledge of ourselves, and it becomes either good or sour inside.
—Pearl Bailey, *The Raw Pearl*

MENTION FEBRUARY 2001 AND DOUG WILL CLAIM IT AS the month I tried to kill him. "Save it for a judge," he will interrupt if I try to speak in my own defense.

When the call came that two snowmachiners were missing in an avalanche, this time near the East Fork of the Matanuska River about a hundred air miles north of Anchorage, Doug was dragging around the house with bronchitis. For years that wouldn't have been nearly enough to stop him; even broken ribs and a newly inflated lung hadn't kept him from pounding up a glacier in an attempt to rescue a man from a crevasse. But of late he was increasingly feeling the weight of all the bodies he had dragged from the mountains. By the time he stopped counting in 1980, he had recovered a hundred of them, stiff and broken—victims of airplane crashes, hypothermia, falls, drownings, and other mountain accidents. By 2001 the number surely exceeded two hundred. Almost fifty had been avalanche fatalities, and at least fifteen of those were friends or acquaintances. The abuse we had taken during the Coyne search had left him asking, "Who needs this?" more loudly and bitterly than ever. In the spring of 2000, at age fifty-four, he turned in his pager, resigned his volunteer position with the Alaska Mountain Rescue Group, received a covey of distinguished lifetime service awards, and began devoting more of his extracurricular time to creating fanciful sculptures from animal bone and wood. Still, when I'd get a call, as often as not Doug would dash out the door with me. Six months after his "retirement," Doug was the first rescuer to reach Nick Coltman, the man paralyzed on Flattop Mountain. For the East Fork avalanche, though, he scurried to help me get ready, shoved sandwiches and a thermos of hot tea into my pack, gave me a kiss, urged me to be careful, and curled up with a book.

Except for the unpardonable fact that it was a killer, the East Fork avalanche was a beauty. It was big and idiosyncratic, venturesome and elegant. Alice in Wonderland would have called it "curiouser and curiouser." It was exactly the kind of avalanche that made Doug and me fall in love with avalanches and, by extension, with each other. It inspired in me the same awe and excitement I'd felt twenty years earlier, on encountering my first avalanche. I wouldn't have been at all startled to have Doug lope up beside me with his bowlegged gait, his beard not yet the color of clouds, and his strong, gesticulating hands painting pictures in the sky. With untarnished delight, we would have crawled over every inch of the avalanche, our fingers extracting stories from the snow and our inclinometers charting the lay of the terrain. Doug, the master of analogy, would have seen not just a thick mound of debris but a pile as high as a four-story building covering three city blocks with snow blocks the size of office desks and pickup trucks. But now I stood without Doug at the receiving end of the avalanche, with a corpse at my feet and another man still missing. That the avalanche had taken the lives of two people was not at all surprising; what was truly incredible was that five others had managed to survive.

Our small group searched gamely that night, but humans and dogs alike had been sickened by fumes from the military helicopter that transported us to the site. Long after Doug had gone to sleep, I called him from a lodge twelve air miles from the accident site, where we'd flown to regroup. He was still untangling himself from his dreams even as I began laying a trap for him, dropping bits of information like cookie crumbs.

Several hours into a hard ride that took them into deeper and deeper folds of the Chugach Mountains, seven snow-machiners had entered a narrow canyon in single file. Blinkered by their helmets, they saw only a low-angled gully leading to a golden splash of sun on the slope above. It was −10 degrees Fahrenheit, and the winter sun couldn't lift high enough in the sky to illuminate or strike the chill from the valley bottoms. All the men wanted to do was to reach the sun, turn off the sleds, and swig warm coffee from their thermoses. With a less restricted view, the gully might have looked as spooky as a dark alley in a tough neighborhood. Bounded by much steeper slopes, it formed a natural avalanche spillway.

As the gully rose to meet the sun, it also steepened. The lead riders had just reached this headwall, with the others strung out below them in a loosely spaced line, when the snowpack shattered. As the slope broke to the east of the men and the spillway began to churn with moving snow, one man was low enough to flee out the bottom of the death trap. The other six attempted to scramble up the velodrome-like west wall and hang on to knuckles of rock even as several of their machines were swept away. Three of these men were snatched into the body of the avalanche; one, improbably, survived. Trapped on his back underneath his flipped machine with his face pressed into the windshield, he was maytagged by blocks weighing hundreds of pounds as he careened more than a half mile and fifteen hundred vertical feet down the mountain at over eighty miles per hour. When the avalanche slowed to a stop, it popped him unceremoniously to the surface. Bruised but otherwise uninjured, he emerged on legs of rubber back into the life he should have lost.

But as I paced the back hall of the lodge talking to Doug on a borrowed cell phone, it was the physical characteristics of the avalanche rather than the human details that I focused upon. The snowpack must have been crackling with energy like a high-tension wire, for once the fracture was initiated, it shot at least three thousand feet around the mountain, wrapping over two ridges and clear-cutting the snow from three separate bowls. Most of the acres of snow set into motion funneled through the gully, though the snow from the farthest shoulder of the mountain had created an entirely separate runout zone a quarter of a mile away. The fracture looked to be only a knee-deep shelf where it had been triggered, but judging from some of the blocks of debris, which reached to more than twice my height, I guessed that in places the fracture had arced into deposits of wind-loaded snow as deep as fifteen feet. The blocks were so hard that pressing my fingers into them with as much force as I could muster barely left an impression.

Doug always greets my estimates with skepticism. During summer rowing trips in the Arctic, if I say the waves are four feet high, he will judge them to be a mere 3.85 feet and accuse me of exaggerating. Now he questioned me closely, but it had been nearly dark when I saw the fracture, first from the canted back window of a helicopter and then from the distant toe of the slide. All I knew for sure was that the avalanche was big, really big, and that because I would be occupied with the search, the specifics would be lost unless I could entice Doug from his sickbed. I promised him that I would arrange for a helicopter to land him beyond the heartbreak; he need only concern himself with the avalanche. When he asked in a hoarse croak, "Are you sure I should come?" I knew I had him hooked.

Mild remorse washed over me the next morning when an Air National Guard Black Hawk helicopter arrived with a new load of searchers, and Doug stepped off looking as though he belonged in a hospital. Skeptically, he eyed the stubby helicopter on inflatable pontoons I waved him over to; Helo One was unavailable, and the Troopers had substituted a helicopter normally used for aerial fish surveys in the flatlands. For nearly the first time in twenty years, we wouldn't be flying to an avalanche scene with Bob Larson, the pilot we'd grown to depend upon as much for his kindness and unmitigated love of Snickers bars as for his imperturbability and competence. But only the month before, at age sixty, Bob had retired from the Alaska State Troopers. Still passionate about flying but tired of being on perpetual call, he had gone to work for a private helicopter company. With a measure of unspoken trepidation, we packed the chopper. Flying with a new pilot in an unfamiliar ship in unforgiving, alien terrain is not altogether different from climbing into bed with a stranger and hoping for the best.

My plan was that the pilot would drop Doug and Scott Horacek, a mountain rescue team member charged with setting up a radio repeater site, as high on the mountain as possible. I would take advantage of the helicopter's altitude to confirm the determination I'd made in dim light that it was safe to search. But at the last minute, the pilot, who was unaccustomed to flying in the mountains and nervous about weight, said that he'd prefer to take only two passengers. I thought myself gallant for bowing out. Doug could take the second look at the fracture, and I'd fly in one of the Black Hawks directly to the debris, along with the first teams of searchers and dogs.

Our helicopter took off about fifteen minutes behind Doug's. I was up front with the National Guard crew so that I could guide them to the site while the rest of the rescuers were caged by a wire partition behind me. I was the only member of our team with a headset, actually a little telephone I held to my ear. To ride in a Black Hawk is to jiggle like a pebble in a barrel; without a headset, intelligible conversation is impossible. The avalanche was just coming into view when a chorus of expletives from the crew erupted over the intercom. I followed their stares out the window, and there was Doug's helicopter in pieces on the ground. It was splattered upside down, with the ungainly floats hanging in the air and part of the rotor skewered into the snow next to the ship. I could see the moving dots of only two people. Where was the third? Where was Doug in his bright red parka? My heart was already kicking dents in my ribs when Scott's reassuringly calm voice purred in my ear: "Trooper helicopter to Black Hawk. We are okay. We are okay. We are all okay." Behind me I sensed the stir of the other searchers and knew that they, too, were counting bodies and coming up one person short. They had no way of hearing Scott's message and, restrained by a double shoulder harness and stuck on the wrong side of the wire wall, I could only hope that they saw my feeble thumbs-up. As we circled slowly, like a dog looking for the perfect place to lie down, I kept straining without success to catch sight of Doug.

When searching for a crashed airplane, I'd been trained, don't look for the plane, because it might not look like a plane anymore. Instead, search for something that doesn't look like everything else around it. I used this method to look for Doug even as I kept expanding the perimeter of my scan well beyond where I thought he should be. Five hundred feet above the crash site, a flash of motion caught my eye and there was Doug, climbing steadily up the gully. It would be hours until we'd be close enough to touch and even longer before I'd hear his story.

Afraid that the floats of the helicopter might turn into skis on snow, the pilot had been uneasy about landing on the thin ridge, which didn't offer much of a perch anyway. He had descended quickly, without circling, into the shadowed valley, intending to land next to the debris. But in the flat early morning light, the valley floor lost all contrast, and when the pilot thought he was still fifty feet above the deck, the right skid bumped a small snowy hill. The helicopter bounced back into the air, still moving forward, but then the skid hit again, this time much harder. Doug, who was sitting up front next to the pilot, says that the horizon began to tilt gradually until it was canted 35 or 40 degrees. Everything took place in slow motion until the rotor caught on the snow, flipping the helicopter with sudden violence. When Doug described the impact to me, he gestured as though he was slam-dunking a basketball or smashing an egg against the ground. The rotor cut loose and the windshield exploded outward into the snow. Although Doug was sitting to the left of the pilot and wearing a double shoulder harness, his head hit the right wall. Less than three seconds after he first suspected they might be in trouble, Doug found himself hanging upside down.

Not long before Doug first joined the Alaska Mountain Rescue Group, some of its members had been injured when the military helicopter they were riding in crashed, and one of the pilots had been cut in half by a flying rotor blade. The threat of being guillotined by the rotor had been dodged this time, but fire was still a worry, especially when the men caught a whiff of leaking fuel. They hurried to get clear of the ship. But Doug, canted diagonally with his full weight on his seat belt, was unable to release the catch. The pilot let himself down until he was standing on the ceiling, and then reached above his head to unfasten Doug's belt. Doug had no choice but to reciprocate ungraciously by dropping down on top of the much smaller man.

Once they were all safely outside, Doug retrieved his hat, glasses, and headset, which had been wrenched from his head and hurled six feet out the open windshield. But there was little else to be done. The helicopter was reduced to scrap metal, and with the temperature well below zero, standing around was going to be an increasingly cold proposition. So within a few minutes of the crash, Doug shouldered his pack and began hiking toward the distant fracture line.

Reassured that the three men were uninjured, I, too, lost no time in bending my energies toward the original mission. By noon the last victim had been found and quarried from ten feet of very hard debris. Though we still hadn't had a chance to talk, by late afternoon Doug and I were at least vibrating hand in hand on a shared Black Hawk ride back to town. The evening passed in the usual blur of media inquiries, and we both sank into sleep almost as soon as we managed to find our way to bed.

At three in the morning, Doug and I jolted awake as though an alarm had just sounded. In a sense it had, for only now, a full eighteen hours later, did the emotional impact of the crash begin to register. Only now, in a dark, warm bed, were we safe enough to let in the full glare of vulnerability. In torrents we began to speak of what might have happened, until there was too much for words alone to say.

In the next few days, Doug would tell his story a dozen times; each telling creating a wider, more comfortable buffer between the past and the future. He wasn't hurt or even unduly upset, but in his usual way, he needed to understand the what and how and why of the accident. The first person he telephoned was Bob Larson. As I listened to Doug's end of the conversation and looked at him cozied up next to the woodstove, I could see Bob just as clearly in my mind's eye, his handlebar mustache planted like a cactus in his leathery face and a sly smile sneaking from its shadow. I knew Bob's questions from Doug's answers, and it felt exactly as though he was maneuvering through the tight space of our living room, testing the wind, and holding a careful hover until it was time to fly in a new direction.

Nine months later Bob Larson was dead. When we heard that a helicopter had crashed in Cook Inlet, killing the pilot and two others within view of our living-room window and a mere half mile from the Anchorage airport, we knew that it couldn't be Bob. This was a routine flight to service the air navigation beacons on a pancake-flat island off the end of the runway, the equivalent of a drive to the corner video store. The total distance was less than five miles, and there was no topography to pose any kind of obstacle. Sure, it was snowing, but Bob had flown thousands of hours in less than perfect visibility, and this time he had the advantage of being able to fly close to the deck. Bob didn't get rattled, even when lunatics shot at his helicopter or his rotors spun inches from unyielding cliffs, and he wasn't a cowboy. Everything was always okay when he was at the controls. He certainly couldn't die on this measly puddle jumper of a flight. The helicopter must have malfunctioned; maybe the rotors had iced up.

Long after the bodies and the helicopter had been fished out of the water and the experts had pored over the minutiae came the verdict. The crash had been caused by pilot error.

Some deaths create a ripple in our daily lives, some a wave of disquiet. For us, Bob Larson's dying threw all the water out of the pond, even though we'd rarely seen him except during rescue missions. The next day we flew out of state on a long-scheduled trip, but no journey was going to take us far enough to escape the sadness or the reckoning that comes when the "unthinkable" happens. If Bob Larson could die in a helicopter, then we could die just as readily doing the avalanche and search work we had long since come to regard as routine. While this certainly wasn't news, we were beginning to listen to a rabble-rousing crowd of doubts. Why were we risking our lives trying to bring back the dead?

We were away during Bob's service, which was more than we thought we could handle anyway. A friend saved the newspaper stories, but Doug has declined to read them. In the *Anchorage Daily News*, the search had made front-page headlines two days in a row, and later in the week, there had been a long feature article in which Bob was remembered as a "helicopter angel," a "real steady Eddie," and a "search-and-rescue miracle worker." Paul Burke—the state's rescue coordinator, who had shared a retirement party with Bob—said, "I think his DNA had gears and oil in it. When he went into a helicopter, he didn't so much fly it as wear it." A half-page photo showed Bob as he appears in my memory, in a royal-blue jumpsuit, his right hand curled around the joystick, the Chugach Mountains looming beyond the bubble of the cockpit.

The coverage given to his death, extensive as it was, seemed trifling for a man who had spent a third of his lifetime protecting and rescuing others. And the articles were uncannily similar to those I'd written in my sleep after Doug's East Fork crash. I'd wake at odd hours with sentences in my head, which is not unusual for me as a writer, but when I'd rush to record them, I'd realize that they were a facsimile, complete with character quotes, of what the newspaper would have published had Doug died that day. From prior copy I knew he'd be called "a pit bull for his cause" and that someone would be sure to say, "This isn't the local big-fish-in-a-small-pond kind of guy."

In reading the articles about Bob, I found myself asking more and more questions of the sort not generally answered in newsprint. Does Judy, his wife, still step out on the back deck of their house every time she hears a helicopter overhead? What does she do in the stilling hours at the end of the day, when Bob would call on his cell phone to tell her he was safely down? I remember those calls; he always made them while the rotors were quieting to a stop, before he climbed out of the ship to secure it for the night. Doug and I knew we should go to Judy, but it was she who had the grace to find us and pull us out of hiding. When she told me that Bob had been reading my new book when he died, and that it was still sitting on the bedside table beneath his reading glasses, my heart might just as well have been a crystal vase cascading to the floor.

❋ THERE IS A tremendous spirit of shared purpose during most rescue missions that can give way to a tendency to blame the rescuers, to second-guess the decisions of both rescuers and victims, or to get angry at the victims for killing themselves. We tend to focus our attention upon the dead, but awareness is growing that those left behind may be imploding with denial, rage, depression, and grief. After rescue efforts involving death or trauma, it has become increasingly common to hold what is called a Critical Incident Stress Debriefing (CISD) to help family members and rescue workers cope with the emotional aftershocks. I've always refused to count the number of bodies I've helped dig from the snow, because it seems akin to collecting scalps. The number, though, must have been pushing forty by the time CISD came into vogue in Alaska's mountain rescue community. Certainly, I've long known that because Doug and I are on call statewide, we have been exposed to more death than most avalanche workers in the Lower 48, where each county tends to tap separate rescue resources. I was offered my first stress debriefing after the 1999 accident that killed six snowmachiners at Turnagain Pass. Even Doug, with a closetful of corpses jammed into his psyche, had never before been extended the opportunity.

Even after it was clear that none of the victims would be recovered alive, the Turnagain Pass mission had taken on a life of its own. Seizing on the accident's accessibility and drama, the media had turned the search into a form of reality TV, and even the governor had dropped in on the action.

On site Doug and I were in charge of four hundred searchers, including National Guard troops; a comment by their commander that I would make a good general had made my father proud. With little sleep, an intensifying blizzard, and the pressure of making a continuous string of rapid-fire decisions over five days, I'd grown so exhausted that at one point I had called myself on the radio—"Jill, this is Jill"— providing a moment of unintended comic relief.

What makes me an effective accident site commander is that I can set my emotions aside and think from another portion of my brain. I wasn't sure what was to be gained by welcoming my feelings into consciousness. I knew that more winter was still to come, though I didn't guess that in the next several weeks, seven more people would die in six separate avalanches. I didn't know that in a two-day stretch, Alaska Airlines would hold a southbound jet on the tarmac for Doug so that he could rush to Cordova, where an equipment operator had just been caught in an avalanche; I would helicopter north to run the search for another buried snowmachiner; and a good friend would use his car to gas himself to death. But as I walked into the CISD, I fought the impulse to turn tail and go grab some uninterrupted sleep.

I entered nervously, not knowing what to expect. We were divided into groups, and my cluster of eight search leaders and three counselors was ushered to the distant end of a windowless corridor. My instinct to escape only intensified when I saw boxes of Kleenex lined up on the tables, and it ballooned to near panic proportions when we were instructed to avail ourselves of the bathrooms if need be, for once the debriefing started, we would be shut into the room for the duration. When the lead facilitator requested that pagers, cell phones, and radios be relinquished to eliminate interruptions, an arsenal of electronics was unholstered, clattering comically into the open. With similar disarmament happening in many of the adjacent rooms, it would have been a bad night to be in need of a mountain rescue.

As soon as the door closes, participants in a CISD take an oath of confidentiality, vowing that nothing said in the room will be repeated. The idea is to create a safe-enough harbor that people are willing to expose their feelings. One person might speak with anguish about handling the dead while another vents the frustration and sense of impotence that comes with reacting to virtually hopeless situations. A CISD can feel like a kinder, gentler form of open-heart surgery rather than a group therapy session. Emotions as vital as blood seep to the surface, and strength is discovered in talking and listening, laughing and crying.

Each participant is given a handout detailing the "Common Signs and Signals of a Stress Reaction," which the counselors emphasize are not signs of craziness or weakness but simply indications that the event may be temporarily overpowering. Four columns list dozens of physical, cognitive, emotional, and behavioral symptoms, including fatigue, poor concentration, heightened or lowered alertness, nightmares, intrusive images, anxiety, guilt, sadness, fear, apprehension, depression, irritability, intense anger, withdrawal, change in usual communications, and inability to rest. I looked at the list, sprayed mental check marks next to most of the critical signs, and thought, *So what else is new?* Then I flashed back to a phone call I'd received from Doug a decade earlier, not long after he had stood on the flank of Tincan Mountain swearing at Todd Frankiewicz's body.

When Doug called from Juneau in southeast Alaska, there was a catch in his voice that seemed more than a vagary of the phone connection spanning the six hundred mountainous miles between us. "I have no idea what happened," he said. He'd spent the day teaching avalanche rescue to a classroomful of highway equipment operators. Doug had lived his subject for so long that the crucial information flowed from him as freely as air, and the act of teaching no longer required particularly conscious effort. He had been showing a video, a corny television reenactment of the rescue of a Wyoming snowplow operator thrown off the road by an avalanche that had smashed the truck's windows and cemented the driver in snow. I knew the video well. I'd used it so many times in teaching similar classes that the mere mention started it spooling through my mind. The video begins with an impeccably coiffed middle-aged man in a suit speaking earnestly about dangers in the "dead of winter." As he walks through a warm studio, his image gives way to falling snowflakes, although his sonorous voice remains.

I comforted Doug. Whenever I watch the video, I cry too, especially when the victim's wife stands aquiver next to the hospital bed where her husband lies without moving, swaddled in warm blankets. My assurances carried little weight with Doug, who knows that I also cry watching the sappiest of McDonald's commercials. "But, Jill," he said, with an unmistakable wobble. "You don't understand. I couldn't stop crying for half an hour."

Doug was raised in the era when strong men didn't cry. It bothers him that he is not as "bulletproof" as he used to be, and that even a radio news story can swamp his eyes with tears. He is a silent crier and always turns away, hoping that I won't notice as he brushes his face dry with his shoulder. Each new, blatantly avoidable accident bruises him and leaves him subdued for days afterward. Sometimes he says he feels adrift in a fog. Early in my rescue career, Doug advised me not to look at the faces of those I dug out, warning that I might not be able to forget them. The faces of the dead occasionally parade through his mind in painful detail. I've tried to heed his counsel, though it isn't always possible to escape the crossfire of a victim's stare when I'm on hands and knees in the bottom of a burial pit or when tussling a 200-pound body into a bag. Doug tends to leap to these tasks in an effort to protect others from the toxic levels of exposure to accidental death he feels he has accrued. When I tell him that I think I'm handling it just fine, he says, "Yeah, that's what I thought too."

Even if I manage to avoid imprinting a victim's face into my memory, my mind is host to a crowd of other images: the wedding ring on the cold, waxy hand of the newlywed; the anguish on a father's face as he identifies his son in the body bag; the friend slumped in the back of the Trooper patrol car; the obituary photo in which the smiling victim is holding his two-year-old daughter. Say the name of the victim or the place—Bull River, Boulder Creek, Powerline Pass—and the images tumble forth. Predictably, new fatalities drag memories of old accidents to the surface. It seems to take less and less to trigger the associated stir of sadness and hurt. Doug and I could barely sit through the movie *Titanic;* the bodies looked too real. After the terrorist attacks of September 11, 2001, Doug was stricken with nearly as much empathy for the young uninitiated college students who volunteered to search the wreckage of the World Trade Center as for the more immediate victims.

It seems that I remember best what I most want to forget. Not long ago I had lunch at the home of my former boss. He'd given me my first job in Alaska, launched me in the avalanche business, backed me unflinchingly, even tried to set me up with his son Bruce, an accomplished mountaineer and geologist. In his living room, a prominently displayed photograph of a man I knew had to be Bruce caught my attention not because it captured him so vividly but because the likeness seemed so poor. Leaning in for a closer view, I realized with a flush of shame that I remembered Bruce better dead than alive. The dominant snapshot in my mind was how he'd looked when we found him facedown under four and a half feet of snow, buried only a few arm's lengths away from a dead friend. The winds that stormy day had been ferocious enough to lift me off my feet, and on the way down the mountain, I was inhibited from flying farther only because I'd been tethered to Bruce's body. In the months following the avalanche, I apprehensively awaited questions from Bruce's father, who was plainly devastated by the loss. It was a conversation he never initiated; I suspect he couldn't stand to know any more.

For every fatality, there is usually a mother or girlfriend or sibling who wants us to explain how and why their loved one died. I've learned to buffer myself from a victim's friends and family the night of an accident, when they are probably not ready for the answers and I may not yet be able to subdue my anger—or, increasingly, brutal lack of surprise—at so unnecessary a rerun. Often bewildered relatives, many of whom live in places that are green and balmy year-round, use questions like flashlights to try to shed light upon the new reality that has been thrust upon them. A mother wants to know how her son could survive a war and die when he was only playing in the mountains. A father says his son is an experienced climber who has "climbed mountains all over north Georgia all his life." A woman nursing a ten-day-old baby wonders how her husband could have given mountaineering priority over family. A man wants to know why his hiking partner didn't try to save herself by rolling over and digging into the slope. A snowboarder asks why he survived and his buddy did not. I was working on this paragraph when a friend called asking permission to give our names to a father whose twenty-two-year-old son was killed climbing a few months ago when a cornice broke and sent him tumbling off a ridge.

Sometimes there are no answers. Sometimes the answers are a stew: *No, it wouldn't have made a difference if he had been wearing a beacon. Yes, they had already made several passes on the slope before it broke, but that is not uncommon. I'm sorry, I just don't know . . .* Sometimes, as messengers, we are hated for our answers. I try to keep poisonous information private, locking it deep inside, though it can feel as corrosive as battery acid: *Your husband had an airspace, and the basket of the ski pole attached to his wrist was visible above the snow surface. He was buried alive for hours and likely would have been saved if his friends had kept their cool and searched rather than leaving to summon help.*

My understanding that these accidents don't need to happen has only made it harder to watch as surviving friends and family are plunged into grief and despair. I am tired of being a target for misplaced anger, and I have grown cynical of the flurry of interest in avalanches following a fatality. After nearly every accident for the last two decades, there has been a cry to resurrect the state's avalanche forecast center. Ultimately, apathy has prevailed, and the sad reality is that even the best forecasting cannot prevent people from killing themselves. Pet agendas are typically advanced after every accident—educators lobby for additional education, rescuers solicit more equipment, quiet sports advocates urge the banning of snowmachines from the backcountry, snowmachiners argue that if more territory was made available to them, they wouldn't be forced to ride in dangerous terrain. Some take comfort in knowing that the victims died "doing what they loved to do," but I have trouble getting past the reality that they needn't have died at all.

Because avalanche accidents tend to cluster into weekends and holidays, Doug and I frequently made the transition from digging to dinner party. For many years we did so with relative ease, unrattled by the incongruity of feeling for an ice mask around the victim's face in the afternoon and then foraging through the hors d'oeuvres only hours later. Often we'd clump into a friend's house a bit more than fashionably late, like returning war heroes in our big boots and waterproof bibs. Inevitably the conversation would turn to us, and we'd hold forth with the details of the accident. Now we are much more likely to skip planned engagements and slip home. We can't stand listening to ourselves trot out the same well-worn explanations and sound bites. It makes Doug feel like a "trained monkey" to respond to the same media inquiries over and over again. In the post-avalanche quarterbacking that always occurs, I've become increasingly intolerant of hearing victims called "stupid" by casual readers of the newspaper or other backcountry travelers, who seem to think themselves immune to error. The victims may have made poorly informed choices, but clearly they headed into the mountains in pursuit of life rather than death.

Doug no longer zestfully chases every lead that might shed a glimmer into Alaska's avalanche history. He's not kidding when he says he feels as though he has lived a couple of hundred years and survived more than four thousand avalanches. His database is a treasure trove, stunning for the sameness of so many of the events. In mining records squirreled away in a dusty basement, Doug found the funeral receipts for the two men killed in 1917 on the same mountain in Hatcher Pass where Keith Coyne died in 1999. It cost $5 to shave one of the victims and $7 to transport both bodies by wagon. VALLEY BURIED IN HUGE SLIDE appeared in the *Valdez Daily Prospector* in April 1912, next to a story announcing the landing of the *Titanic* survivors. Similar headlines crop up year after year, almost word for word, in all of Alaska's newspapers. Many of those same headlined valleys are now pockmarked with houses, accidents just waiting to happen. No matter how hard Doug has worked to catch up with Alaska's avalanche past, it has been impossible to keep up with the present and, presumably, the future. Now if a tip comes in about a recent human-triggered avalanche, Doug is as likely to let it molder on his desk as to quickly track down those involved and record their story.

We have spent our careers chipping away at a metaphorical iceberg, trying to prevent accidents. Every time we look up, the iceberg appears larger and more treacherous than ever, and we wonder if we have accomplished anything. Though we've taken encouragement from testimonials urging us to "continue to fix kids like me so we don't end up doing something stupid and dead," the bodies are easier to account for than the unknown number of lives we have saved through education. Sometimes someone crosses our path whom we instinctively shy away from getting to know too well. Midconversation with a young, particularly brash outdoor enthusiast whose athletic ability to ski or snowmachine or climb in hazardous terrain far outstrips his ability to assess potential danger, I'll realize that I am picturing him exactly as he will look when he is cold and dead.

❄ AS I WRITE this chapter, another snowmachiner has just been killed in an avalanche at Arctic Man, the second such fatality in four years. For one weekend every spring, the Arctic Man Ski & Sno Go Classic transforms a lonely stretch of two-lane frost-heaved highway through the Hoodoo Mountains southeast of Fairbanks into Alaska's fourth largest and most lawless city. With the closest town twelve miles away and little more than what writer Beryl Markham would have called "a word under a tin roof," upward of twelve thousand celebrants bring their own lodging and provide their own entertainment. Arctic Man is a motorized Wild West, with a 4,000-square-foot beer tent open until two in the morning and at least three thousand motor homes and campers. The crowds flock not just to watch a race in which skiers and snowboarders are towed at breakneck speeds behind snowmachines, but to ride with abandon. On brightly colored steeds that can cost more than $10,000, they ride everywhere—buzzing across treeless expanses of tundra, flying over bumps, and throttling into the mountains ahead of contrails of blue smoke. Almost everyone hoots or speaks loudly at Arctic Man, for there is no other way to be heard over the perpetual whine of straining horsepower.

Four years earlier, on April 8, 2000, Bob Larson had picked me up in the usual spot at the top of our driveway to fly the several hundred miles north to Arctic Man. Knowing that the invasion was under way and suspecting that the snow was as unstable in the foothills of the Alaska Range as it was elsewhere, the night before I'd stuffed a few chocolate bars and a freshly filled water bottle into my rescue pack and gone to bed alongside my pager. The first fatality occurred while I was sleeping—a twenty-six-year-old man well stoked with liquid intelligence had driven his snowmachine through the encampment at excessive speed and slammed into a parked trailer. The ambulance service would respond to twenty-one other injuries at Arctic Man that weekend, including a broken back and a fractured pelvis. But the call for me didn't come until Saturday afternoon. A snowmachiner had been buried in an avalanche. Worried about the remaining hazard, the Alaska State Troopers on site had ordered the search suspended until the danger could be properly evaluated.

I was waiting for the helicopter when, as an afterthought, I ran to the basement to grab an extra climbing rope I might need to rappel into the starting zone. With the rope tucked under my arm, I turned to leave, but the door, which had been missing the inside knob for a few years, had shut behind me. I'd been stuck in the basement before and had to pound on the ceiling with a broom handle until a chuckling Doug took pity upon me. Now Doug was gone, and the helicopter would be touching down any minute. I could picture Bob quietly puzzling my absence in his unflappable manner, and I could imagine the headline: AVALANCHE SEARCHER COULDN'T FIND WAY OUT OF HER OWN BASEMENT.

I heard the change in the pitch of the rotors that meant the helicopter had landed. Cursing Doug—for I have deemed it his job to fix things like doorknobs—I forced myself to slow down, dropped the rope, pinched the naked little neck of exposed screw with the desperate grasp of a drowning person, and twisted my hand to the right. Free, I ran full tilt up the driveway and was still breathless when Bob greeted me with his time-honored "We have to stop meeting like this."

Actually, we'd been seeing a lot of each other. For me, the winter had been one of spent emotion and adrenaline. It had begun in October with the avalanche death of mountaineering legend, avalanche forecaster, and friend Alex Lowe in Nepal. The death of Keith Coyne and the fury of his family followed in December. Next came the millennium avalanches of January and February, which had trashed power lines, buried Jerry LeMaster in Cordova, and killed Kerry Brookman along the Seward Highway. And only the day before, a woman who had taken several of our workshops had been caught and killed in an avalanche while skiing.

William Coty, the forty-three-year-old snowmachiner missing at Arctic Man, had triggered an avalanche earlier in the day that had buried him to his waist. Normally, troopers in their street shoes, gold-striped pants, and holstered hips stick to their patrol cars, but with avalanches releasing everywhere, they were desperate to forestall disaster. Two officers had borrowed hot-rod snowmachines and met Coty at the bottom of the slide, where they urged him at least to wear an avalanche beacon if he was going to do more riding. Coty, shaken and wet, assured the troopers that he intended to be careful. But he ventured out again, still without a beacon, and a few hours later triggered a second slide. This time he wasn't lucky enough to swim to the surface.

I could see Coty's story carved like graffiti into the mountainside. He had shot upward of dozens of other highmark tracks, veering away from the windblown ridge where they were concentrated and into an untracked bowl. He had reached his apex just below a corniced ridge, and the trajectory of his descent had carried him left into the middle of the slope. There the avalanche commandeered the story, erasing any evidence of Coty. From afar it looked as though a heart had been drawn in the snow, with Coty's last-ever highmark track forming the right side and the fracture line sketching in the left. The fracture itself was beautiful, a clean surgical cut ranging from ten inches to five feet deep and shimmering golden in the last of the day's sun.

The Troopers had made a good call in suspending the search, for the whole right-hand side of the slope where Coty had made his turn had not yet avalanched. To evaluate the snow, I walked off the edge of the ridge with my inclinometer in one hand and probe pole in the other. Almost right away, both confirmed what I had already surmised. The unreleased portion of slope was 6 degrees gentler and had not been loaded by the wind. It was safe to search.

I wanted to assess a short, shadowed hill lurking above one side of the debris, so I radioed Bob that I was going to walk down the path. In the meantime I asked him to ferry another mountain rescue team member named Michael Thompson in from the road, along with a witness who might be able to give us a fix on where Coty was last seen. As I descended, though I zigzagged and looked carefully, I could find no signs of Coty or telltale pieces of gear. Searchers earlier in the day had removed Coty's helmet and snowmachine without leaving any markers, effectively wiping the slate clean.

I am accustomed to family members doing what I would do in such a situation, hovering close, demanding to know what actions are being taken, clamoring to help, and urging speed. So I'd been surprised that no family or friends had pushed forward when we'd first landed at the Arctic Man encampment and was dumbfounded when Bob returned with Michael on board but no witness. Rumor had it that Coty's five companions had gone back out riding; whether true or not, none were to be found.

Most victims are buried within two hundred feet of their machines, usually within a smaller radius of forty feet. If I have nothing else to go on, I will begin by searching just upslope of the machine, which, because it is larger and heavier, tends to get carried slightly farther. But now, in the dark, almost seven hours after the avalanche, with no witnesses or clues to help pinpoint the most likely search area, no search dogs yet available, and only Michael and me on site to probe the long ribbon of debris, there was little point in searching. The best use of time was to head back to camp and mobilize the necessary resources to mount an effective search in the morning.

My first thought when I walked into the big camper that was serving as Arctic Man headquarters was that I'd never seen so many troopers mustered to deal with an avalanche accident. Then, with stories of slashed tires on the patrol cars, the true reason for such a show of force began to register. I needed to make several calls, so an officer drove me out to the highway, where there was better reception. The three-mile access road was jammed with campers and snowmachine trailers parked end to end, set aglow by bonfires and surrounded by arsenals of steak-laden barbecue grills, convivially arranged lawn chairs, and coolers chocked with beer. Some of the most alcohol-emboldened were jumping over eight-foot-wide rings of waist-high flames on their snowmachines, launching themselves over the bonfires from ramps built of snow.

The troopers arranged for Michael and me to lay our sleeping bags down in the medic tent, strategically located between their own camper and the Arctic Man security tent, and a few flaps down from the beer hall. The aura of lawlessness was so pervasive that a sergeant offered to escort me to the outhouse, fifteen yards away. Space in the medic tent was so tight that when I lay on my pad on the hard plywood floor, my feet doubled as doorstops. After jamming in earplugs and finding a crude semblance of sleep over the constant rev of passing snowmachines and the bellow of the beer tent, I dreamed that a woman was standing over me, crying. As I surfaced into unwanted consciousness, I realized that there really was a woman standing astride my feet, talking tearfully to a medic. She was the first of three women to falter in that night saying that she had just been raped.

In the morning I wanted nothing more fervently than to find Coty and flee. Two handlers had driven through the night from Anchorage with their search dogs, and Coty's brother and other companions were now on hand to tell us what they knew. In short order we found Coty forty feet from where his helmet had been picked up. He was lying faceup, underneath four and a half feet of snow. There was a look of astonishment upon his face.

The newspaper headlines announcing Coty's death were interchangeable with those reporting a fatality at Arctic Man four years later: HIGHMARKING TRIGGERS DEADLY SLIDE. By now you've read enough to anticipate the details. Beautiful day. Recent wind. Fresh avalanches. Three men. Narrow, steep-sided canyon. High-powered machines. One man stuck on a slope. Two others climb to help him. Avalanche catches all three, but only completely buries one. No avalanche beacons. Of the fifteen snowmachiners in the area who stop to assist with the search, only two are equipped with probes and shovels. One survivor tells a *Fairbanks Daily News-Miner* reporter that the possibility of avalanche danger "crossed their minds . . . but you never think it's going to happen to you." The victim was twenty-four years old and engaged. His father tells the paper that "he could say a million things about him."

I look at the photographs of the accident site on the Web and inadvertently let out a groan loud enough to wake the

dog. Enough clues to have prevented the accident are evident even from one photograph. There are no surprises, only
profound relief that I didn't have to be summoned to search
for the body.

❄ TOWARD THE END of a Critical Incident Stress Debriefing, the lead mediator always asks if any good can come of
the experience. The question is generally followed by a long,
shuffling silence. When others do begin to speak, I am still
thinking. I am not searching for things to say, however, but
only for a way to sort them. It astonishes me, really, how
much good I can think of.

For me, the most obvious answer is that we are here,
gathered around a table in support of each other. I look
around at the wind-chapped faces, and there is so much I
know about each one and so much they know about me.
We've been drawn here for a mix of reasons. Some have no
motive other than a true desire to help others. Some love to
be needed; some get a rise from the pager on their hip and
the helicopter rides. Some are rarely separated from a hat or
jacket bedecked with rescue patches; some have desk jobs
and see rescues as a chance to spend time in the mountains.

I am a specialist. I do few rescues that are not avalanche-
related and seldom respond to a search for a lost child or a
climber stuck on a cliff. Initially I was drawn by the avalanches themselves. As an educator and forecaster, I needed
to understand as much as I could about how and why and
where people were getting caught. Rescues were about what
I could learn, which doesn't make me much of an altruist.
And absurd as it sounds, as a lover of avalanches, I've even
felt some responsibility to mop up after their messes. I never
harbored much hope that I could save the victims. More and

more, however, I believe in my ability to keep rescuers alive. In the room are people who welcomed me when I knew little and now trust me with their lives.

The rescue and avalanche business has made me a zealot about this trust thing. It frightens me that the people in the room have invested me with so much credibility that they never question my "go" or "no go" decisions. This means that I'd better be damn sure I'm right, which means presuming little and not being afraid to admit what I don't know. In the mountains I count on my partners, as they count on me, to do exactly what has been promised; safety is predicated on dependability. This premium on reliability has washed over into my everyday life, making me fanatical about keeping my word and meeting deadlines, and causing me to lose an inordinate amount of trust in those who let their pledges slide.

Selfish as it sounds, one good result of body recoveries is that they remind us of the thin line we tread. One minute a skier is skimming through powder like a bird; the next she is dead, with her arms outstretched and her leg twisted around her head. Reminders of our vulnerability can cause us to freeze in alarm, afraid to leave the kitchen table, or, like the bonds we have to one another, they can affirm life. As one man confided in a soft voice while we finished digging his friend from an eleven-foot-deep hole, "I just keep thinking that I'm glad it isn't me." Sometimes amid the distractions of daily existence, we lose track of how simple our choices really are: we can live or we can die. For me, death has added conviction to life.

In retrospect, I think the local mountain rescue community should have held a Critical Incident Stress Debriefing after Bob Larson's death. Even though it did not fall to us to fish his body out of the gray drink of Cook Inlet, his loss cratered into our souls. It took me four months to write a condolence letter to Bob's wife, and the loutish delay was not for lack of trying. As I drafted this chapter, I called several members of the mountain rescue team to talk about Bob. Each was quick to call me back, but when I broached my subject, to a person each postponed the conversation. Only one has called again as promised. We are trained to know that talking will help us heal, but our hearts are still too bruised and swollen to let the words escape.

Helped by Doug's fair warning to avoid looking at the victims' faces and to vent my emotions, I have not been as haunted by death and déjà vu as he, but this may be nothing more than fewer years of exposure and a lingering naïveté. Regardless, we are bonded in spirit, so watching his pain has been enough to make it my own.

For Doug, Bob's death was a catalyst. Avalanches and mountain rescue have been a passionate crusade, but like a wounded warrior, Doug could feel his strength for the fight weakening. He wanted to make room for other pursuits while there was still space in his life. Cruising magazines began to replace the stacks of avalanche literature in the to-read pile next to his side of the bed. In kayaks and rowing boats, we've paddled the equivalent of the circumference of the earth, but our muscles are unlikely to hold up for another round. We began to talk about finding a big boat on which to put our little boats, a floating home in the wild.

We're not Sisyphus. We can stop rolling the rock uphill whenever we choose. Blinded by a sense of obligation, even of ownership, Doug and I forgot that for a while. Sometimes at our most discouraged, we can also forget that the past is a resource, not a destiny.

CHAPTER 11

Truth or Consequences

*Looking down the barrel of the Behrends Avenue avalanche path,
circa 1972, Juneau*   TOM LAURENT COLLECTION

Reportedly a transcript of a radio conversation between a
U.S. naval ship and Canadian authorities off the coast of
Newfoundland, released October 1995.

AMERICANS: Please divert your course 15 degrees to the north
to avoid a collision.

CANADIANS: Please divert your course 15 degrees to the south
to avoid a collision.

AMERICANS: This is a captain of a U.S. Navy ship. I say again,
divert your course.

CANADIANS: No, I say again, you divert your course.

AMERICANS: THIS IS THE AIRCRAFT CARRIER USS
MISSOURI. WE ARE A LARGE WARSHIP OF THE U.S.
NAVY. DIVERT YOUR COURSE NOW.

CANADIANS: This is a lighthouse. Your call.

IT IS NOT CLEAR THAT JERRY LEMASTER, THE SURVIVOR OF
the January 2000 avalanche that ripped apart the 5.5 Mile
neighborhood in Cordova, feels lucky. He lost his partner,
his house, his belongings, and a measure of his health. When
I went to see him two weeks after he'd been thumped back
to life, he had only slightly more color in his hollow cheeks
and a livid surgical scar that zippered the length of his left
arm. "Today," he said from bed, "if folks told me that chute
could load up that much, I wouldn't believe them. But you
won't see me out there rebuilding."

Even once he was physically able, Jerry refused to revisit
the wreckage of his former life. He wasn't alone in his dread.
Mark Kirko, the first to make contact with his friend and fellow firefighter through the small opening in the debris, was
more unnerved by the avalanche than by working the gory
scenes of car crashes. Since it is impossible to venture more
than 5.5 miles by road from town or even to reach Cordova's

airport without transiting the neighborhood, he couldn't avoid the area forever, but the first few times he was driven through, he kept his eyes squeezed shut.

Conversely, Bob Plumb, Cordova's sole paid firefighter, returned obsessively. Shielded by a video camera, he captured footage of the immediate aftermath, including the skeleton of a log house so caked in spindrift from the powder blast that it looked like a freezer relic overdue for defrosting. The house was more or less functional once the trees had been lifted from the roof and new glass put in the windows. As less lucky homeowners salvaged what they could, the beeps of reversing backhoes added a monotonous sound track. The sweep of Plumb's camera caught a mounted grizzly bear head sitting forlornly on the snow and a basketball hoop standing incongruously alone. If anything, the neighborhood looked even more devastated as spring peeled back the veil of snow cover and Plumb fastidiously documented the emerging jumble of crushed cars and boats, building fragments, and sodden personal effects.

Cordova, with its sea otters, rain forests, glaciated peaks, and legendary runs of Copper River red salmon, may be uniquely wild, but the town is far from special in its skirmishes with nature. The most "natural" thing about natural disasters is that they occur around the world with stunning regularity. A quick survey of natural disasters in 2004 alone includes a cyclone in Madagascar in March, landslides in Kyrgyzstan in April, an earthquake in Iran in May, floods in Macedonia in June, and mud slides in Nicaragua in July. Habitually, humans forget to make accommodations for nature in our plans. Though lax at preventing the disasters, we tend to be masters at recovery. We pick ourselves up from nature's knockdown punches with such pluck and speed that often there is no discussion about whether readying ourselves for another round makes sense.

As I write this, the earthquake-triggered tsunami in the Indian Ocean—with the death toll currently estimated at 150,000 and still escalating—is only two weeks old. From the outset, there was talk of how long it would take to "return to normal." Virtually all the damage and deaths occurred within seven hundred yards of the sea, in an area with a history of major tsunamis every few hundred years. Doug and I have spent our careers beating our heads against the same brick wall as a scientist I heard speaking on National Public Radio. He had devoted much of the previous year trying to educate people in Sumatra about what to do in the event of a tsunami but had found disconcertingly few who were willing to pay any attention at all. Though major tsunamis are known to have occurred in 1524, 1762, 1819, 1847, 1881, 1883 (caused by the eruption of Krakatoa), 1941, 1945, and 1977, and island villagers speak of a tsunami in 1907 that killed thousands, to most the risk was impossibly obscure compared to more pressing economic issues. "Does this mean," asked the interviewer, deftly missing the point, "that you predicted the tsunami?" Absolutely not, the scientist replied, reiterating the simple fact that tsunamis of similar or even greater size will occur in the future, just as they have in the past. Such catastrophes are as old as human history, though each time we react with unmitigated shock. Even though it has been only fourteen years since a cyclone-induced tsunami killed 138,000 people in Bangladesh (1991), it has barely been mentioned in discussions of the most recent "unprecedented" disaster. "Civilization exists by geologic consent," wrote historian Will Durant, "subject to change without notice."

Much of the media commentary about the Indian Ocean crisis has centered on the tragedy that might have been prevented if an early warning system had been in place. But a warning system is only as good as the options for escape—even if a warning had been communicated, many living barely above sea level had nowhere to run. Nor does the fact that a warning has been issued mean that it will be heard—or heeded. The most authoritative warning—the ominous drop in the level of the water prior to the tsunami—largely went ignored. (Interestingly, an isolated aboriginal population living attuned to nature on an island off the coast of India managed to survive without losing a single life, apparently because they understood the import of the clues. Similarly, a ten-year-old British schoolgirl who had studied tsunamis in a geography class two weeks earlier saved the lives of her family and about a hundred tourists on a beach in Phuket, Thailand.)

Here's what is likely to happen if millions of dollars are spent putting a warning system in place: During the rest of the century, smaller earthquakes will evoke tsunami warnings, which will precipitate panicky evacuations, especially if they occur within the next few years. But if earthquakes occur without tsunamis and little damage is observed, over time the warnings will trigger less and less concern, until complacency has again taken dangerous hold. This is not an argument against warning systems, but a plea to recognize that in the rush to assign blame and create "fixes," we often overlook the fundamental problem.

The staggering number of deaths in the Indian Ocean makes it impossible to fathom each of the victims as an individual, with distinct needs and dreams. Distance also serves to soften the reality. The scale at which I have dealt with the consequences of clashes between man and nature is small enough that I can still clearly distinguish each of the avalanche victims within my heart and mind. My intimate view has taught me to be less judgmental of individuals, recognizing in their missteps and blindnesses the potential for my own. I find that as I follow the news from the Indian Ocean, I recognize the look of despair in the eyes of both the survivors and the rescuers. I am awash in grief—and yet I cannot be astonished. Even as I have become more personally forgiving, the repetition of tragedies that are more the result of myopic human perspective than the complexities of nature has hardened my skepticism about the possibility of effecting lasting change in society's intransigence toward the natural world.

But then, every so often, something happens to remind me not to let knowledge or cynicism get in the way of hope.

❄ DOUG FIRST TOOK NOTE of the potential threat at 5.5 Mile in 1981 when he taught an avalanche workshop in Cordova. Over the next twenty years, when business brought him to town, he sought out residents and city officials to talk about the hazard. But even as additional construction exacerbated the problem and more people, including renters, began to occupy the neighborhood year-round, Doug couldn't stir up much interest in what seemed a remote and speculative concern. In the thirty years since the land had been parceled into lots and sold for development, the biggest avalanches had done little more than ding up a few buildings and knock over some trees. Most of the residents thought they had seen what avalanches could do at 5.5 Mile. Jerry LeMaster had lived there for almost a decade. But we're destined to lose any waiting game with nature, which has infinitely more time than we do. Thirty years is nothing but a nap in the life of an avalanche path.

The January 2000 avalanche was both a wake-up call and a conversation starter. Jerry had barely been in the hospital long enough to start eating the food before the residents and local government began debating the fate of the neighborhood. Monitoring the discussion from home, Doug and I were convinced the debate would be brief. As the shock subsided and the snow melted, Cordova would rally to put the neighborhood at 5.5 Mile back together again.

Many Alaskans were drawn to the periphery of the continent precisely for its culture of personal freedom, and government restrictions on where one can live or which improvements will be allowed on privately owned property are commonly regarded as provocation. Two days after the avalanche, Cordova mayor Ed Zeine told the *Anchorage Daily News,* "We're looking to possibly condemning that area so that we can be sure no other buildings go up there." Merle Hanson, the owner of an undamaged house near one edge of the path, told the *Cordova Times,* "Some of us fear this zoning plan more than the avalanches." Their attention now fully engaged, Cordova's political leaders called us with questions. Just how bad was the problem at 5.5 Mile? How frequently could catastrophic avalanches be expected, and which properties were at risk? What could be done to reduce the hazard?

❄ ANY AVALANCHE SLEUTH worthy of the title employs a variety of techniques to glean information about the history and capabilities of a given avalanche path. Doug flew to Cordova to start building a case. He began by searching for eyewitnesses to the past, which meant going door to door interviewing the residents of 5.5 Mile and highway workers about avalanches they'd seen over the years. Where doors no longer stood, he tried to track down the displaced homeowners.

The challenge of panning for nuggets of useful data in the coarse gravel of memory brings to mind the words of Welsh writer Dylan Thomas: "I can never remember whether it snowed for six days and six nights when I was twelve, or whether it snowed for twelve days and twelve nights when I was six." Memories are filtered not only by the stream of time but are presifted by what was noticed in the first place. The majority of avalanches in the 5.5 Mile Path had gone unobserved because they stopped short of the road and the neighborhood and were therefore outside the scope of normal interest. Most often useful information is assembled rather than found. One person might remember an event, another might be able to add a date, and a third might recall something about the avalanche's dimensions.

A friend of ours passed along a tip from Richard Davis, one of Cordova's longest residents, then in his eighties. Davis remembered that when he was about five—sometime during the early 1920s—the old Copper River and Northwestern Railroad tracks along the existing Copper River Highway were buried by large snowslides three years in a row. As a consequence, Davis said, the tracks were shifted several hundred yards away from the mountain into the area eventually occupied by the 5.5 Mile houses. Avalanches were only one of the hazards with which the short-lived railway had to contend—the enterprise was given such slim odds that an early nickname was "Can't Run and Never Will." Doug has been searching for the railroad's daily train sheets for years, but they have vanished. Newspapers are typically his next best source.

If Doug wants to find dirt on Cordova, he looks in Valdez newspapers, and vice versa. In the early 1900s, when the two coastal towns vied to become the railhead to the interior, their rivalry was fierce enough to lead to a shoot-out. Cordova ultimately prevailed, but today neither town has a railroad. On page 2 of the January 16, 1916, edition of the *Valdez Miner,* Doug found a reference to the Cordova Power Company continuing to "have trouble with its transmission lines being carried away by snowslides." He couldn't find mention of the avalanches Davis remembered, but starting about 1910, given the hunger for outside news after the advent of wire service, Alaska's newspapers often gave much shorter shrift to local happenings. In Doug's view they became "much more homogenous and boring." But the newspapers did discuss the especially heavy snowfalls of the early 1920s, which may be as close to corroborating Davis's childhood memory as Doug will ever get.

Doug found another fleck of gold when his attention was drawn to a 1995 book called *Iron Rails to Alaskan Copper,* which included previously unpublished photos found in the attic of a woman whose grandfather had ventured north as a railroad engineer at the turn of the century. Doug stirred when he spotted a black-and-white photo captioned "Gravel pit at Eyak Lake"—not because he has an abiding interest in gravel, but because he could see from the surrounding topography that the photo was taken at 5.5 Mile. By tracking down the elderly author, he was able to establish the year as circa 1907. The cleared area in the center of the photo was exactly where houses would be built sixty-some years later. More interesting was the timber near the lake. The trees were clearly missing limbs on their uphill side, their trunks as bare as flagpoles to a height of about thirty feet. Strong prevailing winds could cause this "flagging," but the area is not subject to such winds. Logging is another possibility, but removing the branches on only one side of standing trees would seem an unusual practice. None of the nearby forests in surrounding areas logged during the same period show similar damage. The culprit was almost certainly a previous avalanche with powder blast forceful enough to strip the trees of their windward branches.

Doug took to the woods on the uphill side of the Copper River Highway to get a better idea of the frequency of large avalanches. Vegetative "trim lines" can be clearly seen in aerial photographs taken from ten thousand feet, with "disaster species" such as alders and willows in the center of the path, small spruce and hemlock trees toward the edges, and old-growth forest beyond the path's boundaries. Doug wandered amid the trees toppled by the 2000 avalanche and, peering through his magnifying glass, began counting rings. The trees' median age was seventy-three years. Had older trees been destroyed by avalanches, or had they been harvested to make ties and buildings during the railroad's construction? All Doug could really deduce was that the 2000 avalanche had been the most destructive avalanche in the path for three-quarters of a century.

Gathering empirical data is so time-consuming that history is often shortchanged in studies of avalanche potential and other natural hazards. But this data is uniquely valuable because it is not theoretical. Through even short snippets, enough of 5.5 Mile's history can be spliced together so that we know that in the last century avalanches have inundated the alignment of the Copper River Highway at least every eleven years. They have overrun the area where the neighborhood currently sits between the highway and the lake an average of every thirty-three years. Even so, the past is not a blueprint for the future. The next big avalanche may not be *as big* as the 2000 event; it may be bigger.

We weighed the history, pored over topographic maps and aerial photographs, dug into snowpack and weather records, and climbed the contours of the 5.5 Mile Path, comparing it to others of similar size and snow climates. After running computer models that projected velocities, impact pressures, and runout distances, we at last came up with what Cordova was impatiently awaiting—a map.

Such a map delineates the avalanche hazard based upon potential impact forces and/or frequency. For land-use planning purposes, the area of avalanche exposure is generally divided into high hazard (red) and moderate hazard (blue) zones. Given this task, no ten avalanche experts will draw exactly the same lines, though the resulting maps should at least look similar. In any case, the lines imply a greater level of precision than is possible, and avalanches will always have the last word. Avalanche frequency and destructive potential obviously tend to decrease with distance away from the mountain or the center of a path. Not all avalanche paths are created with equal destructive potential, however. In granddaddies with big starting areas, deep snowpacks, and dramatic vertical drops—such as Cordova's 5.5 Mile Path—the red zone is extensive because the major avalanches hit with so much punch. Even in the blue zones, it is possible to have houses destroyed or pushed from their foundations, walls sucked out, roofs caved in, vans turned into convertibles, and people killed. Maps can fairly accurately define the exposure to potential danger; then it is up to the community to choose the level of risk they are willing to live with or die for.

In North America we tend to draw the red zone boundaries to represent the limits of "ten-year" events, while blue zones define the estimated limits of the design magnitude (i.e., "hundred-year") avalanche. When the television news leads with images of huddled residents awaiting rescue from the islands their rooftops have become and the reporter speaks dramatically of a "hundred-year flood," many take comfort in the assumption that a flood of the same magnitude won't recur for a century. In reality, though, "hundred-year" events can occur in consecutive years or can skip centuries. By definition, they are expected within a period of more than thirty years and less than three hundred years, which gives them a 1 percent statistical probability of occurring in any given year. With enough time, extraordinarily rare, exceptionally large avalanches are likely to extend beyond the design limits of the "hundred-year" avalanche. The probabilities are small enough to be disregarded when planning facilities unless, for example, the proposed structure is a nuclear power plant with an anticipated life of five hundred years.

Had we been commissioned to identify the hazard zones at 5.5 Mile before the "hundred-year" event in January 2000, our map would have differed little from what we produced in the slide's aftermath. It would, however, almost surely have drawn a howl of protest from the community. At best, the residents would have thought us daft for extending the red zone across the Copper River Highway, past Jerry LeMaster's house, across the loop road, and all the way to the lakeshore. They would likely have hooted to see the blue zone line drawn more than three hundred feet out into Eyak Lake. But with the damage so raw and memories still in focus, no one was laughing in the spring of 2000. As it was, the owners of the houses on the lake side of the loop road had only to look across the street to understand what might have happened to them. The debate centered instead on what should be done.

———

❀ FOR PRECEDENT, Cordova could look to Europe, which has led the world in building avalanche protection. Explosive control, also known as active control, is not generally done in avalanche paths that have houses built at the bottom because, with damage inevitable and evacuation difficult, few want to take responsibility for pulling the trigger or dropping the bombs. The earliest known reference to passive control or defensive protection was written in 1323 by a French forestry inspector who warned that if the trees helping to anchor the snow above a village continued to be cut, avalanches would rain down on the houses. In the 1300s local authorities in Switzerland began creating *Bännwalder*— woods from which the removal of green or dead timber was forbidden. The Swiss village of Leukerbad, tucked at the end of a spectacular mountain valley and renowned as early as the sixteenth century for its curative waters, was among the first to muster a defense. After an avalanche killed sixty-one villagers in 1518, the town built an angled rock wall intended to deflect avalanches away from the houses and barns below. Given the exigency, a bishop in 1756 granted villagers in Switzerland's Rhône Valley permission to do such work on Sundays. By the 1800s diversion berms and triangular-shaped splitting wedges, some built on the uphill sides of churches, had become more common in the Alps. Some towns tried to prevent avalanches from starting by digging trenches and terracing the slopes high on the mountain.

In modern times the 1951 "Winter of Terror" claimed 1,489 buildings and ninety-eight people in Switzerland, while more than a hundred individuals were killed in Austria. Switzerland, where avalanches are the leading cause of death by natural hazard, does not permit new construction in red zones and, where feasible, requires defenses to be engineered for buildings in blue zones. Moreover, with their long and sobering historical record, the Swiss have chosen to use the "three hundred–year" avalanche as the limit of the blue zone. It is impossible to drive through the country without noticing the more than three hundred miles of closely spaced fences and nets that have been built to hold the snow in starting zones, making the high mountain slopes look spiked with combs. The cost of such supporting structures—roughly $1.3 million per acre—is as steep as the slopes they are trying to keep from avalanching. More than $1 billion has been spent over the last fifty years, transforming the Alps into an enviable candy store of mitigation measures. Though some of these defenses were overrun when February 1999 roared in with exceptionally heavy snowfall, producing multiple avalanches that filled catchment dams, the death toll was markedly lower than it would have been otherwise. The number of people killed by avalanche inside Swiss buildings during 1951 was eighty-two; during the equally intense winter of 1999, the toll was whittled to eleven.

Staying closer to home, Cordova could take its example from ski communities in the western United States such as Alta, Utah, and Ketchum, Idaho, which have used zoning ordinances to try to mandate humans and mountains into a more peaceful coexistence. Or for a more laissez-faire approach, the town could look southeast to Juneau, Alaska's capital city.

❄ CORDOVA IS about the same size as Juneau was in 1898 when stampeder George Hazelet made a brief stop as he steamed north with the Gold Rush tide. In his diary he wrote:

> "This is a town of from 2,000 to 3,000 people although they claim 5,000. Built at foot of two mountains, hardly

room for the houses, in fact many of the houses are built
on piles and over the tidewater, while others are perched
high and dry on the side of the mountain. It looks as if
the mountain would tumble down and cover the town."

A flatlander from Nebraska, Hazelet was no avalanche expert and no prophet, but Juneau's exposure is not subtle.

In 1982 Ed LaChapelle—a lean, precise thinker and one
of the world's most venerable avalanche authorities—told
National Geographic magazine that "Juneau probably has the
greatest danger of suffering an avalanche disaster as any city
in the United States." Ten avalanche paths spill off 3,576-
foot-high Mount Juneau toward town. The largest of
these—comparable in size to Cordova's 5.5 Mile Path—has
within its scope forty-two houses, a hotel, two major roads,
and part of a harbor sheltering five hundred boats. The high
school is just outside of the path, though students coming
and going are exposed to potential danger. Since 1898 avalanches in this Behrends Avenue Path have run all the way to
tidewater five times. In a maritime climate, heavy snowfalls
are common at higher elevations. On average, two feet of
new snow in the starting zone of the Behrends Avenue Path
adds a quarter million cubic yards of snow to the slope.
From the top of the path, the buildings and boats below look
like bowling pins. They couldn't be clearer targets if they had
bull's-eyes painted on them.

Roughly a half mile north of Behrends Avenue, four separate chutes impact White Subdivision, which we have
unimaginatively nicknamed White Death Subdivision. In
the late 1970s Juneau's building official tried to ban new
construction in White Subdivision but was overruled by the
city assembly on the grounds that people should be allowed
to live where they wished. Planner Steve Gilbertson remem-

bers: "To have a resident pleading to the Assembly to let him build in a known avalanche path was like having a delayed death certificate signed. I've seen snow piled twenty feet tall next to the house they built. That resident is gone now but someone in the future will not be so lucky." The problem, of course, is that property has been sold to buyers who are unaware of the danger, and with development a growing number of residents, children, guests, delivery people, and repairmen have been placed at risk. Ultimately, rescuers will also be exposed to the hazard. Nor is involvement limited to those who visit the site, as taxpayers' dollars get pulled into the mix when the inevitable disaster strikes.

Dorothy Tow, a be-sweatered retiree whose voice rises to a frantic pitch when she speaks of avalanches, is trapped in White Subdivision. Tow avers bitterly that when she applied for a permit to build a small house with the bulk of her savings in the early 1980s, the city did not inform her that she was building in a known avalanche hazard area. During its first winter, Tow's house was hit and damaged by two avalanches within three months of each other. The second avalanche buried her house so deeply that Tow could not see out the first-floor windows. Beyond vexed, she refused to let the firemen evacuate her by ladder, insisting instead that they dig out the long flight of stairs leading to her front door. As of this writing, the house has been hit three times, and avalanches have stopped within a hundred feet of the back wall on at least four other occasions. Tow cannot sell the house, and without selling it, says she cannot afford to leave. Forced to live there, she may well die there.

South of the Behrends Avenue Path, the powder blasts of avalanches breaking twelve feet deep in Chop Gully have swept through downtown Juneau, damaging the city's aqueduct water supply, breaking windows, and sending people running for cover. In a frightening series of photographs snapped one January from a mile away, the town completely disappears in powder cloud. Farther south twenty-one paths threaten five-mile-long Thane Road, creating a gauntlet that residents and the school bus must run every winter day. One of the larger and more active of these paths produces slides that can carry branches and a dusting of snow more than a half mile across Gastineau Channel to the shores of Douglas Island. During the height of Prohibition, a bootlegger who successfully hid from the law five hundred feet above Thane Road was not as lucky in evading avalanches. He was found dead, knocked facedown in his still, in a path Doug named Bootleg.

Since Hazelet's visit, more than seventy-two buildings have been hit, damaged, or destroyed within a ten-mile radius of Juneau. Ed LaChapelle has calculated that a house built in the Behrends Avenue Path has a 96 percent probability of being hit within forty years. The response of Juneau's mayor, William Overstreet, was also quoted in the 1982 *National Geographic* article: "I think life would be awful boring if it were 100 percent safe." Overstreet further opined, "It is better to live a week in Juneau than an eternity in Anchorage." The lead in a *Southeast Alaska Empire* article was "Who is Edward LaChapelle and why is he saying those things about Juneau?"

Before 1946 only an abandoned World War I vintage "pest house" for quarantined smallpox victims stood at the bottom of the Behrends Avenue Path, where small, brightly colored wood-frame houses are now nosed tightly together. In 1962 an avalanche moving close to 150 miles per hour slammed into the sleeping neighborhood at five in the morning. Some residents thought it was an earthquake; others were sure they were hearing the boom of an explosion as trees and cars joined them in their bedrooms. Thirty-five houses on three streets were damaged, and only the splintered trunks remained of almost ten acres of mature spruce and hemlock forest. Two extraordinary strokes of luck befell the residents. First, no one was killed. Second, and more amazingly, there was enough discussion of "freak wind" in the newspaper and elsewhere that the insurance companies caved to pressure and, in a farce similar to the tale of the Emperor's new clothes, chose not to call the avalanche an avalanche. By attributing the powder blast damage to "wind," they allowed the repairs to be covered by policies not liable for "acts of God." Rebuilding began within a day. In weeks, with power poles reinstalled and houses sporting new siding, windows, roofs, and chimneys, almost all physical signs of the avalanche's impact on the neighborhood were gone.

Another slide hit a corner of the neighborhood in 1985. It was about four times smaller than it could have been, but one resident with a backyard full of snow told the *Juneau Empire,* "I think this one was kind of a fluke . . . I don't expect to see this happen again." The avalanche just missed the high-hazard zone home of the man who three years earlier had told the *National Geographic* writer, "If it comes down and wipes us out, then it will be too late to worry." Another resident said, "At least we have some big trees to protect us." These trees, which have filled in since the 1962 avalanche, screen the mountain from view, and apparently from mind.

In the late 1990s, when the path had been quiet for over a decade, Doug and I walked down Behrends Avenue knocking on doors with a NOVA crew filming a documentary about avalanches. Thinking we might get more candid opinions if we remained anonymous, Doug and I didn't introduce ourselves. The homeowners were happy to talk. They told the crew that "the people who live here love it. It is so convenient and cozy." Sipping coffee, one woman said amiably, "I don't know if I'm in denial or what. The last time anything happened here was thirty years ago." This sentiment was repeated many times, in many ways. "This house has been here since 1950, and it is still here," said another woman. Doug broke his cover after the camera had been turned off and offered to show her photographs of damage from the 1962 slide, but she was not interested. An elderly woman who did flip through the pictures showing snow in her living room and her bedroom open to the sky just shook her head and wished aloud that she hadn't seen them. Another resident said, "If you want to know about the avalanches that supposedly happen here, you should talk to that couple in Anchorage—Doug and Jill somebody. They're the ones who can't stop talking about Behrends." A father standing on his deck, the mountain looming over him, said, "It's not nice to have the sword of Damocles over your head, but we were more than delighted to get this house." Footage in the same NOVA program shows the aftermath of an avalanche that destroyed seventeen houses and killed twenty people in Flateyri, Iceland, in 1995. Speaking quietly, one woman said, "We wouldn't have believed it if someone had told us that this could happen." The denials and justifications sound the same in any language.

Juneau has commissioned no fewer than nine avalanche hazard studies, completed in 1949, 1962, 1963, 1965, 1967, 1968, 1970, 1972, and 1992. The latest study, done by Doug and me and an avalanche engineer from Colorado named Art Mears, draws essentially the same conclusions as the first. The paths are ticking time bombs. It isn't a question of whether houses in the exposed neighborhoods will be demolished and people killed, but simply how soon and how devastating the toll will be. Based on historical averages, the Behrends Avenue Path is "overdue" for a slide with consequences impossible to ignore.

I can see the blame-pointing fingers and hear the shocked "How could this have happened?" tone of the headlines in my sleep. I suspect that our pent-up frustration will be no match for the outrage of the injured parties. I wonder at the enormous gulf between our levels of concern, even as I understand that the probability of disaster seems much lower to an individual homeowner concerned with only one house, in one avalanche path, over a relatively short period of time. Because few of the homeowners have had direct experience with avalanches, to them the danger is purely conceptual, while to Doug and me it couldn't be more real. When disaster strikes, the people who scream the loudest for relief are likely to be those who have been the most apathetic about the "so-called" hazard. One might even be the lawyer living in White Subdivision who opposed the city's 1989 proposal to include hazard information on plat maps, telling the *Juneau Empire*, "I don't think I could ever sell my house if a notice goes to the recorder." (Though, in 1987, the city officially adopted the hazard boundaries of the 1972 study, the homeowners prevailed in preventing the recorder's office from flagging the affected properties.) Another might be the man who has lived near Behrends Avenue since 1968 and indignantly told the *Anchorage Daily News* in 2000, "The only danger I have is those agitating avalanche experts who come up from California trying to get a job and giving all the warnings about avalanche dangers." When accepting the

Nobel Prize for literature, Czeslaw Milosz said, "In a room where people unanimously maintain a conspiracy of silence, one word of truth sounds like a pistol shot."

❋ CIRCA 1995 we received several large, stout cardboard boxes in the mail. Inside were stacks of neatly arranged files, and on top was a short note written in a hand made tremulous by Parkinson's disease. "I'm sending these records to you," the note read, "because I think that otherwise they might end up in the garbage. I hope you can put them to good use and wish you luck."

The sender was Keith Hart, who knew that he was months away from death. After working as an avalanche technician on the Seward Highway in the late 1950s and early 1960s, Hart had moved to Juneau with the express purpose of solving the Behrends Avenue avalanche problem. (Never, by the way, did he live in California.) On black-and-white photographs showing the wide scar of the Behrends Avenue Path, he and his network of observers drew in red ink the outlines of slides that released each winter. During the winter of 1971–72 alone, they recorded fifty-nine avalanches. Most of these stopped far short of the houses and thus would surely have gone unheeded if not for his observations. On the back of each photograph, he and his team carefully noted the contributing weather factors. Hart wrote the 1967 and 1968 hazard reports and was the voice of reason at endless city council meetings throughout the 1970s. The boxes Hart sent to us included these meticulous records as well as a copy of a letter he had written to Ed LaChapelle in 1971: "It gets lonesome crying wolf and frightening when the wolf is looking in the window when no one hears you."

In 1971, after Hart issued a real-time warning of dangerous avalanche conditions in the Behrends Avenue Path, real estate developer Ron Maas told the *Southeast Alaska Empire,* "There's been altogether too much publicity on this avalanche thing." William Macomber, owner of the Breakwater Inn at the bottom of the path, agreed and worried about spooking those with lodging reservations. In 1973 Macomber became mayor of Juneau. In 1975, after reviewing the latest hazard analysis's recommendations—which included restricting development in high-hazard zones— Mayor Macomber told the *Southeast Alaska Empire* that implementation "would be aiding and abetting cutting our throats. I don't see any value in complicating our fight to survive." At the time he was defending Juneau against a campaign to move the capital to a location more central to the bulk of the state's population.

The option of doing nothing to diminish a hazard is often the cheapest and easiest, and therefore the most attractive. In some areas where the cost of mitigation exceeds long-term repair and replacement costs, it can be the option of choice. In Juneau, however, the "do nothing" option is destined to be the most expensive. To paraphrase U.S. historian Thomas Bailey, every time history repeats itself, the price goes up.

In 1996, fifty years after the problem first began to be discussed in Juneau, the League of Women Voters conducted a survey of homeowners in the Behrends Avenue Path. Of those who responded, 50 percent had never seen a copy of any of the hazard studies, 25 percent were not aware of the hazard classification when they bought, 31 percent planned to add on to their homes (and were unaware of the restrictions on doing so), 46 percent intended to sell within five years, and 81 percent had no difficulty securing mortgages or insurance policies. The League of Women Voters survey also examined the assumption that property values drop when attention is focused upon avalanche or other geophysical hazards. After the 1972 study, land values in the Behrends Avenue Path doubled, and home values rose by 10 percent. After the 1982 *National Geographic* article tagged Behrends Avenue as the nation's worst avalanche disaster risk, both land and building values jumped as Juneau prevailed in the capital move fight. Immediately after an avalanche hit one house and barely missed several others in 1985, house values increased by roughly $7,000. The survey noted that though regulations in Sun Valley, Idaho, require notice of avalanche hazard on deeds and in all property ads, mandate special engineering on new construction, and discontinue public ambulance and school bus services during periods of high risk, property values have soared. In Vail, Colorado, which is the largest ski area in North America and draws millions of visitors a year, site-specific studies must be done for each parcel of exposed property. Even marginal land is expensive, and many of the affected homes carry seven-digit price tags.

The myriad of studies have presented Juneau with similar mitigation options ranging from forecasting, evacuation, and rescue plans to the construction of structural defenses to buying out the exposed houses or offering land swaps. Moving buildings and people out of the high-hazard zone is inevitably the preferred recommendation. Ed LaChapelle has said, "When protecting fixed installations from avalanches, never depend on repeated acts of human intervention over a long period of time. Sooner or later, somebody will screw up." Some city planners have acknowledged that moving homes out of danger zones is the city's best option, and in the League of Women Voters survey, 88 percent of the Behrends Avenue respondents said they would be willing to trade property or be bought out by the city. Such a land swap was made with Anchorage homeowners after the 1964 Great Alaska Earthquake registering 9.2 on the Richter scale (the 2004 Indian Ocean earthquake was a 9.0) demolished their houses. The old properties were so clearly worthless at the time that the titles were never rescinded. But over the next twenty years, the land scabbed over, vegetation masked the scars, and memories faded. The empty lots, still underlaid by untrustworthy, unconsolidated clay, crept back on the market, and many now sport the fanciest, most spacious homes in Anchorage. The shortness of our memory for natural hazards is borne out by purchases of earthquake or flood insurance, which tend to spike after a major event and then lapse steadily after only a few dormant years.

In Juneau limited private land and tightening budgets combined with ignorance and apathy have forged iron handcuffs. The town limps through each winter with no formal forecasting program, no evacuation procedures to minimize the threat during periods of high hazard, and a little-practiced rescue plan. After rancorous debate, a "buyers beware" ordinance was defeated in the municipal assembly, though theoretically homeowners and real estate agents are required to notify under state disclosure laws. Construction in high-hazard zones is permitted, though it is limited to single-family homes and new buildings are supposed to be engineered to withstand avalanche impact. The costs of such special engineering along with a scarcity of vacant lots and prohibitions on new subdivisions have combined to dissuade construction. No new homes have been built in the high-hazard zones of either the White or Behrends Avenue

Paths since 1987. Theoretically, bedroom additions to existing homes are disallowed, but this is difficult to enforce. No incentive has been provided for the owners of exposed homes to protect their houses. In many cases it would be difficult to do so, either because of the anticipated destructive force or because the defenses would only divert the snow onto neighboring properties.

"We need a buyout," says Bill Glude, a local avalanche specialist who has devoted hundreds of hours to the problem, "but what we get are Band-Aid on cancer measures." In a January 6, 2005, editorial to the *Juneau Empire,* Glude was called "Chicken Little" and a "nickel and dime operator" by the irate owners of a Behrends Avenue property, a house that lost its roof in the 1962 avalanche. Having owned their house without incident since 1993, they asked, "Isn't it about time for all the Juneau alarmists to quit actively trying to devalue our property?" When opportunities for disaster planning or federal appropriations have arisen, the city has requested funds for a buyout, but has been consistently rejected. In 1996 planners proposed developing an incentive program for city acquisition of avalanche-exposed houses, but pointing to financial constraints and the range of similarly affected properties in other hazard areas such as floodplains, the local assembly declined this proposal.

The statements Doug and Bill and I make to newspapers today are virtually identical to Hart's. He died in 1996 knowing that the problem outlived him, and I suspect that the three of us (who have also never lived in California) will do the same. At the beginning of every season, Doug and I wonder if this is the fated winter. We used to wish ourselves near when the big Behrends slide finally hit. Now we hope to be far away and unavailable for comment. The devastation, which will be on a scale far greater than Cordova's, will speak for itself. Bill Glude says, "The problem really isn't any different than alcoholism. Everyone knows there's an elephant in the middle of the living room, and everyone pretends it's not there."

❊ WHEN THE ELEPHANT at 5.5 Mile finally trashed the living room and kicked the walls to splinters, Cordova ironically had to seek help from Juneau. The city of Cordova submitted a local declaration of disaster to the seat of the state government. In quick succession, so did a number of other communities hard hit by the Millennium Avalanche Cycle and storms. The collective cry led the governor of Alaska to issue a Declaration of Disaster Emergency. Normally, that is the end of the line, but for the first time in the history of the United States, the force of avalanches billowed all the way to Washington, D.C. On February 17, 2000, President Clinton issued a national disaster declaration. That piece of paper sent the cavalry into Alaska.

The cavalry didn't ride horses, but they demonstrated startling speed and agility, especially from a government organization with the regimental-sounding name Federal Emergency Management Agency (FEMA). The agency traces its beginning to a fire in New Hampshire big enough to spark help in the form of the Congressional Act of 1803. At the time of the Cordova avalanche, FEMA was an independent agency but has since been folded into the new Department of Homeland Security and charged with managing recovery from disasters precipitated by terrorists as well as by nature. In a sense, it must now respond to a broader range of "acts of God."

The cavalry also came prepared to spend money. Earthquakes, hurricanes, wildfires, floods, and even volcanoes have given FEMA more than ample practice, but never before had the agency responded to an avalanche. Borrowing a strategy designed to encourage people to move out of floodplains, FEMA made a one-time, short-fuse "buyout" offer to Cordova and Valdez, which also had a neighborhood hit by an avalanche during the Millennium Cycle. Each town would be given a slim months-long window in which to write and pass a zoning ordinance that included a "no build" proviso for high-hazard avalanche zones. Only once this ordinance was in place would FEMA, working in conjunction with the Alaska Department of Emergency Services, purchase red-zone properties (at pre-avalanche prices) and pay the costs of moving intact red-zone houses to new locations and hooking up utilities. FEMA was adamant that a second buyout offer would not be tendered if and when another avalanche occurred.

The first, and what we assumed would be the last step in the process, was the hazard mapping, which Doug and I completed for both communities. We'd seen the contentiousness of zoning firsthand in the 1980s when, faced with a proliferation of development in avalanche danger areas, we helped try to persuade the Municipality of Anchorage to adopt an ordinance focusing on the identification of problem areas and buyer notification. It wasn't particularly restrictive, but the residents' cries of protest were loud enough to kill it. Ironically, in the absence of an ordinance, banks and insurance companies have sometimes applied a stricter standard, and many of the same residents have come to wish for the guidance of zoning regulations. Still, the trend of building on marginal land exposed to avalanche hazard has gone largely unchecked in the United States, and over time we will see an attendant rise in deaths and damages.

If some 5.5 Mile property owners were initially chilly to the idea of passing a restrictive zoning ordinance, residents in the affected Valdez neighborhood were much hotter. They had experienced a "ten-year" event rather than the "hundred-year" stunner in Cordova, and many still didn't understand the scope of their problem. At a public meeting, Doug tried to close this knowledge gap with graphic photographs, and the hardest-hit residents—including those with holes still in their buildings—spoke in voices that carried. By May both communities had adopted tough ordinances largely unprecedented not only in Alaska but in the nation. Even more unusual, each of the dozen or so red-zone residents in both towns decided to retreat in the face of hazard, and even a few of the blue-zone home owners in Valdez relocated.

By summer the runout zone at the bottom of the 5.5 Mile Path was beginning to look empty. The house that had been sliced apart while Christal Czarnecki fought for balance in the shower sat on blocks ready to be moved. The wreckage from Jerry LeMaster's house and the other demolished buildings had been bulldozed into piles and carted away with the twisted boats and cars. Some of the structures too damaged or too worthless to move were set ablaze so firefighters could practice for the next disaster.

In ceding at 5.5 Mile, Cordova didn't lose a battle—it ended a war and gave itself a new future. There is talk around town, of course, about what should be done with the high-hazard land on which building is banned in perpetuity. A summertime picnic area, a campground, a bike trail, and a gravel storage site have all been suggested. I've heard rumor of a cemetery. Though it seems irreverent to brand a graveyard an attractive nuisance, there would be great irony in breaking with history only to put the living at risk while visiting the dead. The most fitting epitaph for 5.5 Mile might be "Here we leave ample space for nature."

❄ EARLY IN MY CAREER, I attended a meeting in which Doug was butting heads with an entrenched bureaucrat who kept insisting that office-bound meteorologists, with no knowledge of snow or mountains, could ably forecast avalanches. "Look," said Doug, rising to leave and gesturing toward the wall of the conference room. "I can tell you that the wall is blue. I can show you that the wall is blue. Other people can confirm that the wall is blue. But if you continue to insist that it is yellow, there is nothing I can do."

Doug and I are subject to our own bouts of denial. We're not as diligent about seeking regular medical care as we should be. And we built our house in an area with winds so fierce that they may make it implode next week, or ten years from now, although of course we like to believe it will stand forever. Still, we both chose to take on the cause of keeping people from getting killed in avalanches, and for many years, joining forces only strengthened our resolve.

Increasingly, though, we have found ourselves asking the question posed by philosopher Søren Kierkegaard: "How did I get into this and this and how do I get out of it again, how does it end?" New victims are born every year. We have taken on a problem that ultimately we cannot solve.

At our avalanche workshops, we like to recite a parable that was masterminded by Bill Glude, with minor contributions by us and others. It begins, "Brethren and sistren, today I want to speak on morality and the snowpack. You see, life is like the snowpack. There are many paths you can choose. You are free to metamorphose as you please." We suggest,

"You could become a shining example of a moral, responsible snowflake . . . develop close bonds with your neighbors, sinter and strengthen the entire social fabric." Or, alternatively, "Forget about your neighbors, go all out for personal growth. Go ahead, indulge yourself, build those facets and striations, build great showy crystals, steal vapor from your neighbors. Yes, you could go all the way and become one of the depth hoars, live in sin and degradation down in the red-light district of the snowpack, down close to Hades."

Our "sermon" always draws a chorus of amens, but beneath the hearty laughter, we are dead serious. Snow is unique as a material because it exists within a degree of its melting point, readily shifting between liquid, solid, and vapor. Its constantly changing structure and strength make it an apt metaphor for our own evolution; its variability also underscores the existence of rhythms other than our own. Snow has taught me to pay attention, to look forward and back, to anticipate, and, above all, to keep evaluating. In doing so, it has taught me something about how to lead my life. "The realization of impermanence," says Tibetan lama Sogyal Rinpoche, "is paradoxically the only thing we can hold onto, perhaps our only lasting possession."

Doug was my most influential mentor when I was getting into avalanches, and in the last few years, he has played just as pivotal a role in teaching me that it is okay to let go of them when I am ready. Much of avalanche and weather forecasting depends upon pattern recognition, but I don't want my own life to become too predictable or narrow. We continue to run the Alaska Mountain Safety Center and to take on projects like evaluating the potential avalanche hazard affecting a proposed new oil-spill response facility in Cordova. But when we passed the Alaska Avalanche School off to longtime instructors in the fall of 2004, I felt less as if I was closing a door than that I was walking into intriguing new rooms of my life.

Our goal is to live as seamlessly as possible with nature, in any season. As I worked on this manuscript, Doug slowly gained latitude as he brought our newly purchased two-masted trawler through the Panama Canal and up the west coast of the United States. I wrote some of these chapters aboard as we made the trip home to Alaska together. When we chugged into Cordova, friends were on hand to greet us—with berries on the bushes, spawning salmon packed into the creeks, bears fat with fish, seals rolling in the sea, and eagles circling the sky, we spent a happy week. Only when we were departing did I realize that this was the first stay in years that hadn't included a drive out to 5.5 Mile. And then, with an exuberant smile, I remembered that there was nothing to see.

We hadn't left town far behind when four black-and-white Dall porpoises began crisscrossing the bow of our boat, swimming with dazzling speed and leaping for no reason other than the pure joy of it. They didn't know where we were headed—and for that matter, neither did we.

Acknowledgments

Do not be misled by the fact that my name alone appears as the author of this work. Telling these stories may be the closest I'll ever come to holding the hearts of others in my hands; this book could never have taken shape without the trust and help of many. I cannot conceive of an order in which to express thanks, so I will simply begin with the person most likely to read these words first. Of course, when she does so, she is apt to tweak them here and there.

I changed publishers in order to stay with my editor, Rebecca Saletan. Though I'm not sure I want her to know just how much influence she has over me, I would follow her anywhere. Maybe all editors are as exquisitely insightful, talented, warm, funny, honest, and astonishingly generous with their time, but I don't think so. Becky will read a draft (one of many) and casually note, "I think that the paragraph on page 12 would do better between the second and third paragraphs on page 231." Typically, I can't even remember what is on either page, but I inevitably find that the errant paragraph slips in as suggested, with no transition needed. What I both hate and love about Becky is that she pushes me beyond the comfortable edge of my ability, and for that I will always be beholden. Thanks are due to many others at Harcourt, including Stacia Decker, Lynn Pierce, Linda Lockowitz, Vaughn Andrews, Jennifer Jackman, and all those involved in aiding and abetting this book's long journey to the shelves. Copy editor Erin DeWitt deserves huge credit for her mind-boggling attention to detail. Stuart Krichevsky is wise in ways of the world that I am not. I am lucky to be able to depend upon him as an agent, staunch advocate, and friend. Thanks as well to his assistants Shana Cohen and Elizabeth Coen. John Glusman and others at Farrar, Straus and Giroux were kind, helpful, and extraordinarily gracious.

The irony is that to write a book about avalanches, I had to hide from them. If my dog, Bodie, and I ever run away from home, we are likely to be found at the seaside sanctuary of Michael, Michelle, and Pepper O'Leary in Cordova, where the light is always changing and sea otters are never far. Or maybe we will be deep in the mountains in a cabin generously loaned by Ed LaChapelle and Meg Hunt. When Ed, approaching age eighty, hauled water, cut firewood, kept my computer humming on solar power, popped popcorn, translated narratives from German, made dinner, and did the dishes for me for weeks on end—all so that I could write about a subject on which he is a definitive world authority— I knew the natural order to be profoundly skewed. If Bodie and I still haven't turned up, there's a good chance we are ensconced in a tiny haven in the woods crafted by Dan and Anne Billman or burrowed into a snug cabin in the shadow of the Alaska Range, courtesy of the Denali Foundation. Thanks are due Bodie himself; he can be counted on to always remind me that there is a day waiting outside.

I can do nothing but bow to those who have shared

painful pieces of their lives. Thanks to Jerry LeMaster, Dewey Whetsell, Bob Plumb, Mark Kirko, Christal Rose, Nick Coltman, Maggie Balean, John Stroud, Skip Repetto, Jenny Zimmerman, Regan Brudie, Jerry Steuer, Judy Larson, Aedene Arthur, and the Coyne family, especially Patsy, Angie, and Wes. Without the gift of their perspective, these stories would have no dimension.

Space precludes listing all those who helped me get the details right, but special thanks are due Paul Brusseau, Paul Burke, Kevin Siegrist, Pat Murphy, Terry Onslow, Dave Hamre, Reid Bahnson, Dave Goldstein, Joe Perkins, Este and Ilysa Parker, Steve Kroschel, Jerry Bell, Wayne Rush, Scott Simmons, Tom Smada, Bruce Tremper, Blase Reardon, Dale Atkins, Nick Parker, Christine Pielmeier, Bill Glude, Don Bachman, Jim Bay, Clair Israelson, François Valla, Steve Casimiro, Steve Gilbertson, Tim Maguire, Tom Laurent, Bob Janes, Craig Lindh, Dorothy Tow, Richard Murphy, and Lynn Hallquist. Thanks also to Marne Lastufka for taking on business logistics so that I could write, and to Nancy Pfeiffer, Blaine Smith, Kip Melling, and Paul Wunnicke for carrying on the crusade of the Alaska Avalanche School.

Without friends and family to encourage, prod, and feed me, I would be lost. It may take me years to reciprocate their efforts, but I will never stop trying. Natalie Phillips has again loyally held my hand throughout the entire writing process, beginning from the blankest of pages. Even during his own struggle to rebound from a debilitating bicycle accident, Mike Davidson has been unwavering in his support. I wrap very grateful arms around Janis Fleischman, Jennifer Johnston, Ellen Toll, Evie Witten, Jerry Lewanski, Randy Hagenstein, Nan Elliot, Jo Fortier, Alisa Carroll, Dave Blanchet, Marj and Ralph Burgard, George Clagett, Carol Jewell, Les and Libby Schnick, Peter Brondz, Barb Maier, Dave Hickok, Madeleine Grant, Hal Egbert, Brad Meiklejohn, Laurie Berg, and John Connolly. My embrace extends across the country to my sisters, Dale Fredston and Susan Fredston-Hermann, and their families. I could not be prouder or more appreciative of Lahde Forbes, Sunna Fesler, and Turi Fesler; my love encircles their spouses, partners, and children.

When it comes to expressing my debt to my parents, Elinor and Arthur Fredston, I am stricken with a more hopeless than usual case of writer's block. I simply cannot summon the words to thank them for a lifetime of unstinting love and generosity. If they do not feel the bottomless depths of my respect, gratitude, and love every day, then I am failing miserably.

So far, Doug Fesler's name has been conspicuously absent from these acknowledgments. That is only because it belongs everywhere. We have shared a passion for avalanches, and for each other. He has been the mentor, friend, partner, taskmaster, husband, soul mate, and muse of my dreams. I have dedicated this work to Doug because, without him, there would be no story.

CPSIA information can be obtained
at www.ICGtesting.com
Printed in the USA
FSOW01n1554011117
40639FS